Lighten Up!

A Practical Guide to Residential Lighting

By Randall Whitehead, IALD

Light Source Publishing
San Francisco, California

Publisher	Light Source
Illustrations	Catherine Ng
Designers	Clifton Lemon
	Rolf Mortenson
Typeface	FF Meta Plus

Printed in the United States of America
10 9 8 7 6 5 4 3 2 1

Library of Congress Cataloging-in-Publication Data

Whitehead, Randall   1954-
        Lighten up!: a practical guide to residential
        lighting / by Randall Whitehead. -- 1st edition
                p. cm.

Light Source Publishing
1246 18th Street
San Francisco, California 94107
Telephone 415 626-1210

Lighten Up!

FORWARD

In the 1990's "home sweet home" has become the place to be, according to a recent Roper Reports survey about the preferences of mainstream Americans. People are spending more time at home, using it to fulfill a greater number of individual and family needs.

Home is a place to socialize, with larger kitchens to accommodate resident and guest cooks, and great rooms for informal parties. More than ever, home is now a place for work: It is estimated that between 20 and 40 million people currently do some type of their work at home, whether in a separate room furnished as an office, or at a desk in the corner of the living room.

People also want "home" to mean a safe place, as well as one where they can enjoy the great outdoors in the evening after a hard day's work by extending entertaining to the patio and garden areas.

Throughout all these varied spaces, lighting is essential to making them both enjoyable and functional. Well-done lighting can transform a glare-filled, shadowy, annoying place into one that is comfortable and flexible, enhancing the occupants' feelings of well-being. Advancements in technology have resulted in the development of specialized tools which simplify installation and help the lighting reveal the best aspects of the environments in which we dwell. The designer will learn how to use these cold, hard tools to visually shape a home to meet flesh-and-blood human needs.

Presented here is a valuable, detailed approach to lighting the home that is unique because it not only explains the how of lighting design but also provides insights into the why behind design choices. Hand-in-hand with details on the technology of light is respect for aesthetics, and the overriding philosophy that lighting is for people. The author, Randall Whitehead, has the advantage of being not only an experience lighting designer but a master communicator and teacher as well. Read on, and revel in bringing lighting to its full potential in the most important place in all of our lives-the home.

Wanda Jankowski
Author of *Designing With Light*, *The Best of Lighting Design*,
and *Lighting Exteriors & Landscapes*.
Former Managing Editor-in-Chief of *Architectural Lighting* magazine

Dedication

I would like to dedicate this book to my family, who learned how to put the "fun" back in dysfunctional.

Acknowledgements

My deepest gratitude to Catherine Ng for the excellent drawings that fill this book. I also thank Tim Brace for his computer expertise and for absorbing my misplaced frustrations.

Judy Anderson, Naomi Miller and Alfredo Zaparolli were my wonderful technical experts, who know how to get an idea across without loading us down with technical jargon. Marian Haworth, a teacher of lighting design, was invaluable as the eyes of a student making sure I got the information across without confusion.

I would also like to thank Clifton Lemon and Rolf Mortenson for turning a plain manuscript into readable art.

Lighten Up!

A Practical Guide to Residential Lighting

Introduction

Enjoying Light

There are three elements within each space that need lighting: art, architecture, and people. Think about lighting the people first - you must humanize the light.

This book is intentionally different from the many books written on lighting design. The emphasis here is on the art of lighting, not just the technical aspects.

Think of this manual as a guide to Applied Aesthetics. Light is an *artistic* medium. What you will learn is how to paint with illumination, using various techniques to add Depth, Dimension, and Drama (the three D's), while at the same time humanizing your environments. Remember, you're not only lighting art and architecture, but the people within the space as well. This book will show you how to create rooms that feel instantly comfortable and are flattering to your clients, with the ability to dazzle at the touch of the button.

Today, all architects, interior designers and other related design professional need to know at least the basics about good lighting design. Nothing you learn will have more impact on your designs than lighting, because illumination is the "straw that stirs the drink". You give every single object or space in a home its appearance, tone, and impression through how you light it. Yet people still add lighting as an afterthought: "Oh, yes, let's also do some lighting", after the architecture has been laid out and construction has begun.

This is a huge mistake. Lighting design needs to be brought in as an integral design element, along with all the other design components at the beginning. The lighting budget, too, should be comparable to the other main design elements of the project. Lighting is not just an option; it can make or break your project.

Homeowners need to learn something about lighting design as well, so they will not be lost when making important decisions about their living spaces. There are many architects, interior designers and contractors who will say they know all about lighting design, but have had little training; the homeowners are often left with poor lighting that works against the overall feeling they desired.

All the new technologies of the past decade have increased lighting's importance. Who needed lighting expertise when the only thing available was a ceiling socket and a light bulb? That's all changed, but people's thinking has not. Fluorescent lighting alone has gone through a revolutionary change, and, with energy considerations and construction codes, is now a must for a home.

People have also changed the way they live and entertain, often congregating in the kitchen or a *great room*. Sometimes the house is designed as an *open plan*, where the kitchen, dining room, living room, and family room all flow together. People tend to move more from space to space, instead of staying in one room. The new entertainment centers and control systems have also changed the way people use their homes. People are cocooning more. Well done lighting design has to take all these needs into account, and you can use the new technical advances to make lighting versatile enough to accommodate everything.

Overview—What to Expect

The first half of this book deals with design tools you will need to put together a well-designed lighting plan.

The second half of this book explains what to do on a room-by-room basis for single-family homes.

Remember, it's the aesthetic approach to lighting design that will be stressed. This book is a springboard to your imagination. The possibilities are limited only by your own creative abilities...and, of course, the client's budget.

Section One

Understanding Light

"Arthur, I think I liked our little nest better before we put in track lighting."

Chapter One

THE FUNCTIONS OF ILLUMINATION

Light performs these basic functions: decorative, accent, task and ambient — the well-integrated layering of the four within each space will create a unified design. "The mark of professionalism in lighting is the absence of glare." - General Electric.

The new technologies and developments in lighting over the last decade have created opportunities for approaches to lighting only dreamed of in the past. Lighting technology has greatly evolved from the times of candles and gaslights, yet many of your clients have not updated their thinking much beyond that. We can now achieve lighting effects that match our virtual reality, techno-magic world. Plus, we can do it within a reasonable budget, without dramatically changing the way we live. At the same time we can increase the comfort and convenience level in our living spaces.

Lighting can be a tremendous force in design; it's the one element that makes all the rest work together. Yet it has for so long been the second-class citizen of the design world, and the results have left many homes drab, uncomfortable and dark. Too often the blame goes elsewhere, when improper lighting is the culprit causing the discomfort. Let's take a leap from nineteenth-century lighting to the next plateau by welcoming new lighting possibilities and techniques, and sending design into a new era of dramatic comfort.

Light has four specific duties: To provide *decorative*, *accent*, *task*, and *ambient* illumination. No single light source can perform all the functions of lighting required for a specific space. Understanding these differences will help you create cohesive designs that integrate illumination into your overall design.

Decorative Light

Luminaries such as chandeliers, candlestick-type wall sconces, and table lamps, work best when they are used to create the sparkle for a room. They alone cannot adequately provide usable illumination for other functions without overpowering the rest of the design aspects of the space.

For example, a dining room illuminated only by the chandelier over the table creates a *glare-bomb* situation. As you crank up the dimmer to provide enough illumination to

The Museum Effect — When art becomes visually more important than people within the space. Even museums now add additional illumination beyond accent light to help reduce eye fatigue by cutting contrast in the overall environment.

see, the intensity of the light causes every other object to fall into secondary importance. The wall color, the art, the carpeting, and especially the people, are eclipsed by the one supernova of uncomfortably bright light. They won't see all the other elements, no matter how beautiful or expertly designed.

By nature, any bright light source in a room or space immediately draws people's attention. In the best designs, the source of ambient light will not be visible.

Similarly, linen shades on table lamps draw too much attention to themselves. Consider using a shade with an opaque liner and perforated metal diffuser fitted on top to direct the illumination downwards over the base, the tabletop and across your lap if you're reading.

Accent Light

Accent light is directed illumination that highlights objects within an environment.

Luminaries such as track and recessed adjustable cans are used to bring attention to art, sculpture, tabletops and plants. Just like any of the four functions, accent light should not be the only source of illumination in a room. If you use only accent light, you end up with the "museum effect", where the art visually takes over the room, while guests fall into darkness. Even museums now add additional illumination beyond accent light to help reduce eye fatigue by cutting contrast in the overall environment.

Subconsciously, the people will feel that the art is more important than they are. Of course, some of your clients may feel that the art *is* more important than the guests: Their desires must be taken into account, even if they seem to be incorrect. Sometimes, you'll be able to compromise on a design that provides *some* ambient light. A guest will just have to try to be witty or profound enough to compete with the art.

Accent lighting thrives on subtlety. A focused beam of light directed at an orchid or highlighting an abstract painting above a primitive chest can create a wondrous effect. People won't notice the light itself. They see only the object being illuminated. The most successful lighting effect achieves its magic through its very invisibility.

In the movies, if we can tell how a special effect has been achieved, we feel cheated. We don't want to know, because we want to think it's *magic*. In lighting, it should be no less the case. We want to see the effects of light, but the method needs to remain unseen, hidden, or an illusion. That subtlety is what will create a cohesive wholeness, allowing the design, the architecture, the furnishings or the landscape to become the focus in the space; not the luminaries or the lamps glaring out from within them.

Task Light

Task light is illumination for performing work-related activities, such as reading, cutting vegetables and sorting laundry.

The optimal task light is located between your head and the work surface. That's why lighting from above isn't a good source of task light, because your head casts a shadow onto your book, computer keyboard, or ransom note.

Overhead lighting or incorrectly-placed task lighting often contributes to the problem of "veiling reflection". This occurs when light comes from the ceiling directly in front of you, hitting the paper at such an angle that the glare is reflected directly into your eyes. This causes eye fatigue. Think of it as the mirror-like reflection of a light source on a shiny surface. The surface may be a magazine page, thermal fax paper, or any visual task that has shiny ink, pencil lead, or any amount of glossiness. The veiling reflection is a brightness that washes out the contrast of the print or picture (see Drawing 1.1).

Another related term is *photo-pigment bleaching*. When you try to read a book or a magazine outside, sometimes the brightness of the page makes it difficult to read. You end up moving to a shaded spot or tilting the magazine until the sun isn't hitting it directly.

A reflective surface is always a reflective surface, which means you can't eliminate glare if you are focusing light onto a mirror-like finish. What you can do is redirect the glare away from the normal viewing angle. That's why a light coming in from one side or both sides, instead of directly overhead, is more effective. It directs the glare away from your eyes.

Portable tabletop luminaries with solid shades often do the best job for casual reading, because they better direct the light and don't visually overpower the room when turned up to the correct intensity for the job at hand. You may be thinking, "Well, that's fine and dandy for some Euro-chic interior, but what about my Louis the Sixteenth library?" Well, a bouillotte lamp (see Drawing 1.2) does a great job of task lighting, as does a banker's lamp (see Drawing 1.3). Fluorescent linear lights are also a good source of task illumination at a desk with a shelf above the work surface or in the kitchen mounted under the overhead cabinets.

Veiling Reflection refers to the glare and eye fatigue resulting from overhead light hitting directly on white paper with black print, as if you were trying to read through a veil.

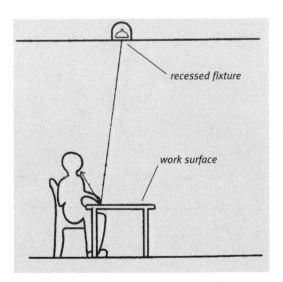

recessed fixture

work surface

Drawing 1.1 (left)
Veiling Reflection (glare) occurs when task lighting is improperly placed directly overhead

Drawing 1.2 (right)
Bouillotte lamp
Here a traditional lamp performs a good job of providing task light

> Ambient light is the soft, general illumination that fills the volume of a room with a glow of light, and softens the shadows on people's faces. It is the most important of the four functions of light, but is often the one element that is left out of the design of a room or space.

As we go from room to room in Section Two, you will get examples of properly placed task lighting.

Ambient Light

Ambient light is the soft, general illumination that fills the volume of a room with a glow of light, and softens the shadows on people's faces. It is the most important of the four functions of light, but is often the one element that is left out of the design of a room or space.

The best ambient light comes from sources that bounce illumination off the ceiling and walls. Such luminaries as opaque-bottom wall sconces, torchieres, indirect pendants and cove lighting can provide a subtle general illumination without drawing attention to themselves. You could call it the *open hearth effect*, where the room seems to be filled with the light of a glowing fire.

Just filling a room with table lamps is not an adequate source of general illumination. The space becomes a lampshade showroom, where the lamp shades are the first thing people see as they enter. Let these portable luminaries be a decorative source, creating little islands of light. Using those opaque shades and perforated metal lids can turn these luminaries into more effective reading lights. Utilizing other sources to provide the necessary ambient light lets the decorative luminaries create the illusion of illuminating the room, without dominating the design.

The inclusion of an ambient light source works only if the ceiling is light in color. A rich aubergine ceiling in a Victorian dining room or a dark wooden ceiling in a cabin retreat would make indirect light sources ineffective, because the dark surfaces absorb most light instead of reflecting it.

A second possibility would be to use a luminaire that essentially provides its own ceiling.

Drawing 1.3 (left)
Banker's lamp
Another traditional type of luminaire that provides good task light

Drawing 1.4 (right)
An RLM fixture coupled with a ceramic bowl reflector lamp provides a wide splay of ambient light without relying on the reflective qualities of the ceiling itself

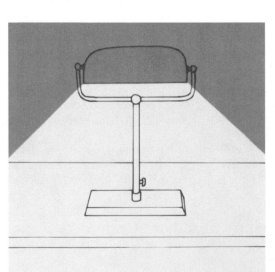

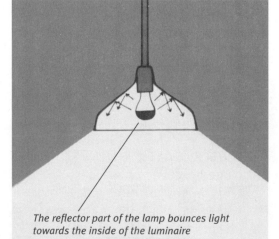

The reflector part of the lamp bounces light towards the inside of the luminaire

One luminaire that has been out on the market for many years is a metal shaded pendent generally known as a RLM pendant. (Drawing 1.4). It has a painted white interior fitted with a silver bowl reflector lamp. The illumination is bounced off the inside of the shade itself, instead of the ceiling, to provide an adequate level of ambient light.

There are more modern versions of the RLM, such as the one shown in Drawing 1.5. The halogen source fitted within an integral reflector bounces light off the dish-shaped reflector and down into the room cavity.

There are many ways of getting ambient light into a room.

Ambient light, just like the other three functions, should not be used by itself. What you end up with is a *cloudy day* or *flat effect*, where everything is of the same value, without depth or dimension. It is only one component of well-designed lighting.

Light Layering

A lighting design is successful when all four functions of light are layered within a room to create a fully usable, adaptive space. Good lighting does not draw attention to itself, but to the other design aspects of the environment.

Once you have a good understanding of the functions of light, you can decide which are needed for a specific area. An entryway, for example, desperately needs ambient and accent light, but may not need any task light, because no work is going to be done in the entry. However, there may be a coat closet, which would need some task-oriented illumination.

What we often see is a house lighted for entertaining only. It has a very dramatic, glitzy look. Many of the design magazines show this type of lighting design. Every vase, painting, sculpture and ashtray glistens in its own pool of illumination. Yet the seating area remains in darkness. What are these people going to do for light when they want to go through the mail, do their taxes, or put a puzzle together with their family? Also, you should know that the design magazines reveal that they often use supplemental lighting specifically for photographing the rooms; those lights won't be there when someone is living in the house, and the effect won't be nearly as wonderful. What it does do is give clients a false sense of what type of illumination downlights alone can provide, which is often all that exists in the space.

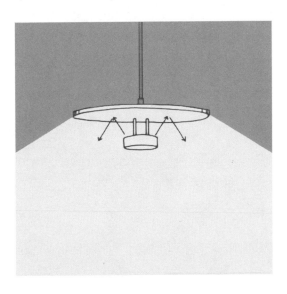

Drawing 1.5
A more modern version of an RLM fixture uses frosted glass or a white metal to bounce the indirect light down into the room.

Please remember that people entertain only part of the time. The rest of the time these rooms are used to do homework, clean and interact with other family members. Highly dramatic lighting is not effective for normal day to day functions.

This doesn't mean that you should eliminate accent lighting. Just don't make it the only option. Simply putting ambient light on one dimmer and accent lighting on another provides a whole range of illumination level settings.

As your clients become more sophisticated about what they want, you should have the knowledge to give them what they want *and* need.

Once the project is finished and someone walks in and says, "Oh, you put in track lighting", it means that the lighting system is dominant. If they walk in and say, "You look great!", or, "Is that a new painting?", then you know the lighting has been successfully integrated into the overall room design.

One solution to this situation is to lighten the color of the ceiling. Sometimes the answer is to alter the environment rather than change the light luminaire. Instead of the whole ceiling being eggplant-colored, how about a wide border in that color with the rest of the ceiling done in a cream color or similar hue? Using a traditional chandelier with a hidden halogen source could compliment the design while adding a '90's sensibility.

A wooden ceiling could be washed with a light colored opaque stain that gives it a more weathered look without taking away from the wood feel itself, as simple painting would do.

Say that your clients are dead-set against changing the color. A luminaire such as the ones shown in drawings 1.4 and 1.5 provide their own reflective surface.

Chapter Two

THE COLOR OF LIGHT—PAINTING WITH ILLUMINATION

Color temperature and color-rendering are two of the most critical aspects in understanding how light and interior design are so very intertwined. Color temperature (chromaticity) technically is as follows: As a piece of metal (black body) is heated up it changes color from red to yellow to white to blue white. The color at any point can be described in terms of the absolute temperature of the metal measured in Kelvin (K). This progression in color can be laid out on a color diagram and now can be specified in Kelvin (GE - Nela Park). This is a tough concept to comprehend.

What does it really mean? The bottom line is this: all lamps emit a color; that color affects the colors you choose in your design. Understanding this interrelationship could dramatically alter how you select your color palette for any project.

Color temperature is a way of describing the degree of whiteness of a light source. Those sources that produce a bluish-white light have a high color temperature and those that produce a yellowish-white light have a low color temperature.

Color Temperature is measured in the Kelvin (K) temperature scale. Understanding color temperature means *unlearning* your concept of temperature in terms of heat. Think of a blacksmith forming a horseshoe. As he heats the metal and it gets hotter it goes from red hot to blue white. When he puts it in water to cool or temper it, it's color changes back to red then to black. Confused? Don't worry. Once you've finished this chapter, you'll have a clearer understanding.

Let's start with the sun. If you were handed a box of crayons and asked to draw the sun, what color would you choose? Most people would choose a deep-yellow. In reality, sunlight is a yellow-white light. It looks yellow compared to the blue sky around it. *Daylight* is blue-white, because it is a combination of sunlight and bluish skylight. In the early and waning hours of the day, there is a golden glow that occurs as the sun passes at an oblique angle to the horizon and is filtered by the atmosphere. But this is not what we

Color temperature is a way of describing the degree of whiteness of a light source. Those sources that produce a bluish-white light have a higher degree color temperature and those that produce a yellowish-white light have a lower degree color temperature.

CRI — Color Rendering Index — A scale that shows how well a light source brings out a true color as compared to daylight.

are talking about in terms of color temperature.

You will hear the term: *daylight fluorescent*, meaning that this lamp comes close to the color temperature of daylight (sunlight + skylight = daylight), which is true. However, the proportions of colors that make up this blue-white light may be significantly different from the proportions in true daylight, so the color-rendering may surprise you. If you expect a warm, sunlight-yellow color of light from these lamps, you will be very disappointed. The color will be much closer to that of cool white fluorescents, as in the greenish glow of a subway terminal, than a lazy summer's day as dusk approaches.

Have you ever gotten dressed before the sun was out or in a closet with no natural light? If so, have you ever looked down later in the day and realized that what you thought was navy blue was really black, or that two reds you chose didn't really go together? You mentally blame yourself for not being careful enough. The truth is that under incandescent light, the colors shift so dramatically that you couldn't tell the difference between navy blue and black, white and yellow, or a blue-red from an orange-red. Trading your incandescent light source for a good color-rendering daylight fluorescent source will allow you to choose your colors more carefully. Please note, though, the color of this light is not particularly flattering - so it's best not to entertain guests in your closet.

Why do we all rush to the window when selecting a carpet sample, paint color or even a sweater? Because we want to see what the *real* color is. This is the correct thing to do — daylight does the best job of color rendering. The quality of light is intense and includes the complete color spectrum, so the material which sunlight hits will be able to reflect its own special hue of the spectrum. All other light sources get rated on their ability to show color hue as compared to the daylight itself. This is called CRI rating (Color Rendering Index). Daylight is the best, so it gets a 100 (a 100% score). The closer the other sources come to daylight, the higher their score in CRI. A 85-90 rating is considered pretty high.

How does this affect your design practices? Looking at your samples in daylight is fine - for daylight situations - but it can be totally inappropriate when choosing colors for night-time or interior settings.

Much of the time designers use incandescent sources (including halogen), that can vary as much as 2200°K from daylight. Under incandescent light, your color selections will shift tremendously. White can go to yellow, red can turn to orange, blue shifts towards green, and grays can turn to tan. Then all your hard work, all your hours of color selection, are for naught.

Take a look at Chart 2.1. It shows how some sample lamps compare in terms of color temperature. At the top of the chart you will see incandescent and fluorescent lamps, which give off a warm (yellower) color. At the bottom of the chart are the more blue-white sources of light, such as daylight and full-spectrum fluorescent, which are very cool colors.

Chart 2.1
*Color Temperature Chart
The lower number in
degrees of Kelvin indicates a
warmer color, the higher the
number in degrees indicates
a cooler color.*

A Sampling of Often Used Lamps and Their Color-Temperature Ratings Measured in degrees of Kelvin (K°)

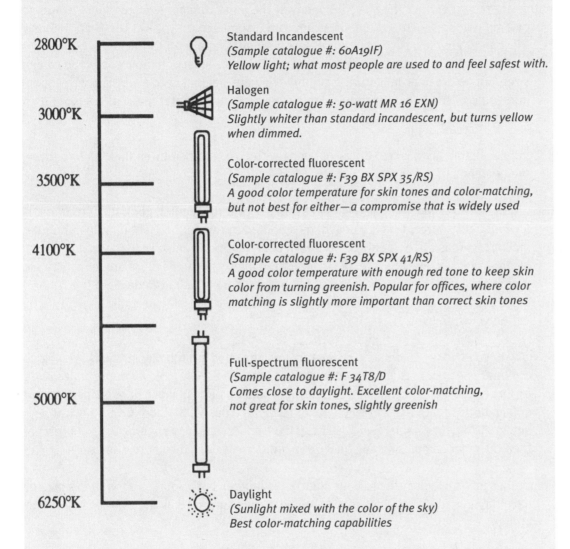

2800°K

Standard Incandescent
*(Sample catalogue #: 60A19IF)
Yellow light; what most people are used to and feel safest with.*

3000°K

Halogen
*(Sample catalogue #: 50-watt MR 16 EXN)
Slightly whiter than standard incandescent, but turns yellow
when dimmed.*

3500°K

Color-corrected fluorescent
*(Sample catalogue #: F39 BX SPX 35/RS)
A good color temperature for skin tones and color-matching,
but not best for either—a compromise that is widely used*

4100°K

Color-corrected fluorescent
*(Sample catalogue #: F39 BX SPX 41/RS)
A good color temperature with enough red tone to keep skin
color from turning greenish. Popular for offices, where color
matching is slightly more important than correct skin tones*

Full-spectrum fluorescent
*(Sample catalogue #: F 34T8/D
Comes close to daylight. Excellent color-matching,
not great for skin tones, slightly greenish*

5000°K

6250°K

Daylight
*(Sunlight mixed with the color of the sky)
Best color-matching capabilities*

Rule of Thumb:
*The lower the number, the warmer the color of light.
The higher the number, the cooler the color of the light.
Reds and peaches look better under a warmer color
temperature, while blues and greens look better under a
cooler color temperature.*

Rule of Thumb — The higher the Kelvin rating, the whiter (cooler) the light; the lower the rating, the more yellow (warmer) the light.

Within these two ends of the chart are many of the most commonly-used lamps. There are colored lamps, such as red or yellow and some colors of H.I.D. (High Intensity Discharge) sources, such as low-pressure sodium, that are off this scale. H.I.D. sources are not normally used in residential design. They are most commonly selected to light large public areas or industrial facilities. They are beginning to find their way though into some of the larger residential landscape projects. Chapter 3 discusses lamp (light bulb) types in greater detail.

Recently available are a large variety of lamps with color temperatures that fall in between the two ends of the chart.

One of the first was halogen (also known as quartz or tungsten halogen). It is promoted as a *white* light source. This is true when compared to regular household bulbs. Standard incandescent is 2800°K, halogen is 200°K cooler (3000°K). Also note that halogen is an incandescent source and like all incandescent lamps becomes more amber as you dim it. It is only whiter when operating at full output. Yet compared to daylight it is 2,000°K more yellow. That's a huge difference. So *white light* is a relative term. Daylight is the definitive white light.

So how do you choose colors for a project that will be used in both daylight and evening situations? The answer is actually very straight-forward. Simply look at your samples under an incandescent source, as well as a daylight source. Choose those hues that are acceptable in both situations. That way, you end up with a design that looks right both at night and in the daytime hours. (See Chart 2.2) If the space you are designing uses fluorescent, then the color samples should be viewed under the selected color temperature.

Remember to show your clients your color board under both sources as well. If you only have daytime meetings, they may be unpleasantly surprised by the way the colors shift at night.

Color-corrected and *daylight* do not mean the same thing. Color-corrected means that the color-rendering of the lamp is good to excellent, intended to compliment the color and/or skin-tone of the people and surfaces within the space. Daylight means the source is close to the color temperature of daylight, but it does not render skin tone pleasantly.

Many lighting showrooms now have *light or color boxes*, which are display cubicles that show a variety of color temperatures and color-rendering abilities from a number of different light sources. This is a great way of seeing how the numerous lamps compare to each other. Light boxes are available for lamp comparisons in most lighting showrooms that work with interior designers.

Rule of Thumb — The higher the Kelvin rating, the whiter (cooler) the light; the lower the rating, the more yellow (warmer) the light.

Lamp	CRI (approx)	Color (approx)	Whiteness	Colors Enhanced	Colors Greyed	Notes
Warm White	52	3000	Yellowish	Orange, Yellow	Red, Blue, Green	—
White	60	3450	Pale Yellow	Orange, Yellow	Red, Blue, Green	—
Cool White	62	4150	White White	Yellow, Orange, Blue	Red	—
Daylight	75	6250	Bluish	Green, Blue	Red, Orange	—
SP30	70	3000	Yellowish	Red, Orange	Deep red, Blue	Rare-earth Phosphors
SPX27	81	2700	Warm Yellow	Red, Orange	Blue	Rare-earth Phosphors simulates Incandescnt
SPX30	82	3000	White (Pinkish)	Red, Orange, Yellow	Deep Red	Rare-earth Phosphors
SPX35	82	3500	Red, Orange, White	Yellow, Green	Deep Red	Rare-earth Phosphors
SPX41	82	4100	White,	All	Deep Red	Rare-earth Phosphors
SPX50	80	5000	White (Bluish)	All	None	Rare-earth Phosphors simulates outdoor daylight
Warm White Delux	77	3025	Yellowish	Yellow, Green	Blue	Simulates Incandescent
Cool White	89	4175	White (Pinkish)	All	None	Simulates outdoor daylight (cloudy day)
Chroma 50	90	5000	White (Bluish)	All	None	Simulates sunlight, sun-sky-clouds
N	90	3700	Pinkish	Red, Orange	Blue	Flatters complexions, meat displays, semi-"cosmetic"
PL	-2	6750	Purplish	Blue, Deep Red	Green, Yellow	Plant/Flower enhancement & growth

Drawing 2.2
A sampling of Lamp Color Specifications
Courtesy of G.E. Lighting, Nela Park

A lightbox is a cube, painted white inside and equipped with different lamp sources that are switched independently. Holding your material swatches or paint chips inside the box under the various lamp sources will help you in the correct lamp decision.

A colorbox is a cube, painted white inside and equipped with different lamp sources that are switched independently. Holding your material swatches or paint chips inside the box under the various lamp sources will help you in the correct lamp decision.

Color Temperature and Plants

Plants love white light. They look lush and healthy under white light sources. Unfortunately, most lights that are used to light plants are incandescent, which, as you now know, has an amber hue. This yellowish light turns the green color muddy and plants look sickly.

What can you do? The answer is to change the color temperature of the light source to a cooler version. Here are some techniques:

Solution One: Have you seen the *grow-light* bulbs that they sell in hardware stores? These are incandescent sources that have been coated to filter out the green-yellow wavelengths emitted by incandescent. Replace the existing lamp with a grow-light version.

Solution Two: Use a color-correcting filter to alter the color temperature of the lamp. They are known as *daylight-blue* filters or *ice-blue* filters, among other names, and come in sizes to fit everything from MR11's to MR16's to H.I.D. luminaries.

Solution Three: Use luminaries that accommodate fluorescent or H.I.D. sources that come in cool color temperatures.

Remember that the sun is white light and the moon is the same, because it is just a reflection of the sun's illumination. So, if you want to create a moonlighting effect with your landscape lighting, here again you need to consider color temperature of the lamps.

Color Temperature and Skin Tone

Even though we grew up with incandescent light and are used to it, it is not the best color for skin, because it adds a slightly sallow cast. A hint of peach or pink in the light is a more flattering light that compliments a wide range of skin tones.

For example, do you or someone you know have a drawer full of makeup that looked great when it was applied in the department store, but when you got home it was just slightly off? Why do you think this happens? Do you just assume you made a bad choice? No, it's not really your fault. When those in-store cosmeticians put you in front of that illuminated mirror, do you know what light source they were using? It's halogen! How often are you seen in a totally halogen environment? Almost never! You are more likely interacting with others in a daylight situation or an evening situation with standard incandescent lights. That's why the colors seem wrong. You need to select makeup under a light source which matches the environment in which you will be seen.

A knowledgeable store will sell you two sets of makeup — one for day wear (the office, the park, boating, etc.), when you are seen in a color temperature between 5000°K and

6250°K, and one for evening wear (the theater, a restaurant, a singles' bar, etc.), when you are most likely seen in a color temperature around 2800°K.

Here are some rules of thumb to follow:
1) Be sure to look at all surface colors - carpets, walls, furniture, etc - under the lamp color specified, as well as daylight.
2) Warm color schemes should use a lower Kelvin source while cool color schemes need a higher Kelvin source.
3) Let your clients see your selections under both lighting conditions, then there will be no surprises or disappointments when the project is installed.

Chapter Three

CHOOSING THE CORRECT LAMPS

When specifying lamps for a project, be sure to stay with the same manufacturer for each type of lamp. That way, the color quality will be consistent from lamp to lamp.

Welcome to the fascinating world of lamps. Lamps are what lay people call light bulbs.

In just the last ten years, a broad spectrum of lamps has evolved for incorporation into our designs. Still, you must understand that there is no one perfect lamp.

Coming to know the various properties of lamps will help you in choosing luminaries, since the choice of lamp will determine your choice of luminaire. After all, it's the lamp that provides that all-important illumination.

You can tell a lot about an incandescent lamp from its name. For example, "75R30" specifies 75 watts. The "R" stands for reflector and "30" tells you that the glass envelope is 30 eighths of an inch in diameter. The first number indicates wattage, the letter(s) indicates shape of the lamp, and the second number is the diameter, measured in eighths of an inch. (See Chart 3.1) A visual sampling of shapes of the most commonly used incandescent, fluorescent, and H.I.D. lamps are shown in Chart 3.2.

Here are some other abbreviations you should know:

A   =   Arbitrary
IF   =   Inside frost
SB =   Silver bowl - The bottom bowl of the lamp is coated to reflect light up.
G   =   Globe - Lamps with a round envelope
T   =   Tubular - Lamps with a tubular envelope
R   =   Reflector - A built in reflective surface.
ER =   Ellipsoidal reflector - A reflector that focus' light in front of the lamp.
PAR =   Parabolic Aluminized reflector - A lamp of heavy glass that controls it's light beam by a reflector and lens
MR =   Mirror reflector - A halogen lamp that uses faceted mirrors to control the light pattern.
S   =   Sign - Lamps for use in signs.

Chart 3.1
Sample excerpt from the
GE Lamp Catalogue

Watts	**75**	—
Order Code	**38207**	*It is important to use this five-digit code when ordering to ensure that you receive the exact product you require*
Description	**75R30FL**	*This information includes the lamp's burning position. It also includes the abbreviations BDTH (Burn lamp in Base Down To Horizontal position) and BU (Burn lamp in Base Up position only). Also shown in this column is packaging information (48PK, Carded, tray, etc.)*
Volts	**120**	*Each lamp's voltage is listed.*
Case Quantity	**24**	*Number of product units packed in a case.*
Filament Design	**CC-6**	*Filaments are designated by a letter combination in which C is a coiled wire filament, CC is a coiled wire that is itself wound into a larger coil, and SR is a straight ribbon filament. Numbers represent the type of filament-support arrangement.*
Maximum Overall Length	**5-3/8 (136.5)**	*Maximum Overall Length in inches and millimeters.*
Light Center Length	**—**	*Distance between the center of the filament and the Light Center Length reference plane, in inches and millimeters*
Rated Average Life Hours	**2000**	*Average Life Hours figure represents the lamp's median value of life expectancy.*
Initial Lumens	**900**	*Initial Lumens figure (in lumens) is based on photometry of lamps operated for a brief period, at rated volts. mean lumens: Also listed in Fluorescent and HID sections.*
Additional Information	**Reflector Flood Inside Frost**	*Typical application and/or other important information.*

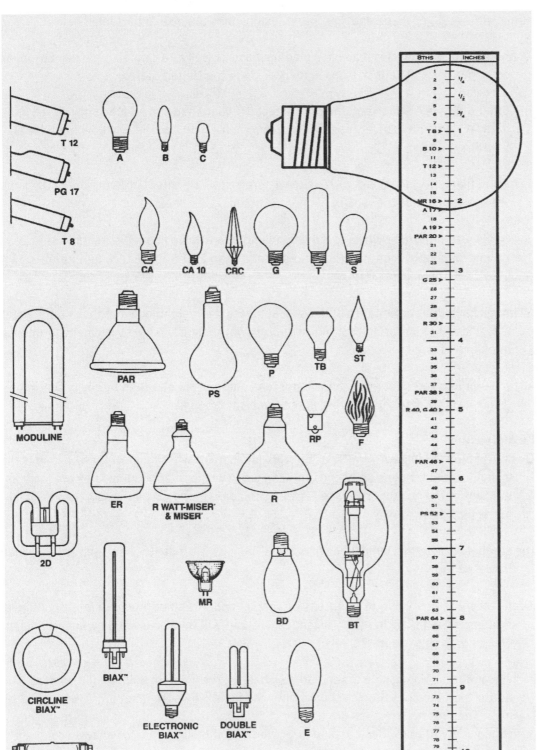

Chart 3.2
*Lamp Diameters and Shapes
All lamp diameters are
expressed in eighths of an
inch. Note these lamps are
not drawn to scale.*

*Courtesy of GE Lighting,
Nela Park*

All lamps fit into three categories: Incandescent, Fluorescent and High Intensity Discharge (H.I.D.).

Along with wattage, shape and size, base designations are also important.

A candle shaped lamp (c) is made in a medium base as well as a candelabra base. European countries use different size bases than we do in the United States.

The other difference is in the method of contact between the lamp and the socket. The standard "A" lamp has a screw base. A MR16 may come with a bi-pin or a bayonet base (see Charts 3.3 - 3.4).

A chart of filaments (see Chart 3.5) is shown here just as a point of information. In the lamp specification guide the type of filament is noted.

Voltage is also important when specifying a lamp. Many lamps come in multiple voltages - 12 volt for small lamps, 110-120 for residential, 130 for rough service and 220-240v for industrial or commercial.

All lamps have their advantages and disadvantages. Understanding what those differences are, and communicating that information to your clients, is a key element to making intelligent design selections.

All lamps fit into three categories: Incandescent, Fluorescent and High Intensity Discharge (H.I.D.). Let's begin with the first of the three categories.

Incandescent
This is the type of lamp with which we are very familiar. It's what we grew up with, and is the visual image that pops up above our head when we get a great idea. They are the least efficient of the 3 categories (see Chart 3.6 to compare how well they do the job of illuminating.)

The standard household lamp, known as an "A" lamp in the industry, has been around since the late 1800's.

Incandescent lamps come in hundreds of shapes and sizes and many different voltages. Halogen sources are in reality incandescent, but will be discussed as a separate lamp type later in this chapter.

In an incandescent lamp, light is generated by heating the filament to illuminate or glow. The hotter the filament becomes the brighter the light. However, lamp life is shortened by heat. Their glass envelopes are usually clear or frosted but can be colored to provide a wide variety of hues. Basic incandescent lamps emit a light that is yellow in color.

When you dim an incandescent lamp, the color becomes more amber, meaning its color temperature becomes lower. This color shift will dramatically affect the look of objects being illuminated. You must take this into consideration when designing a space.

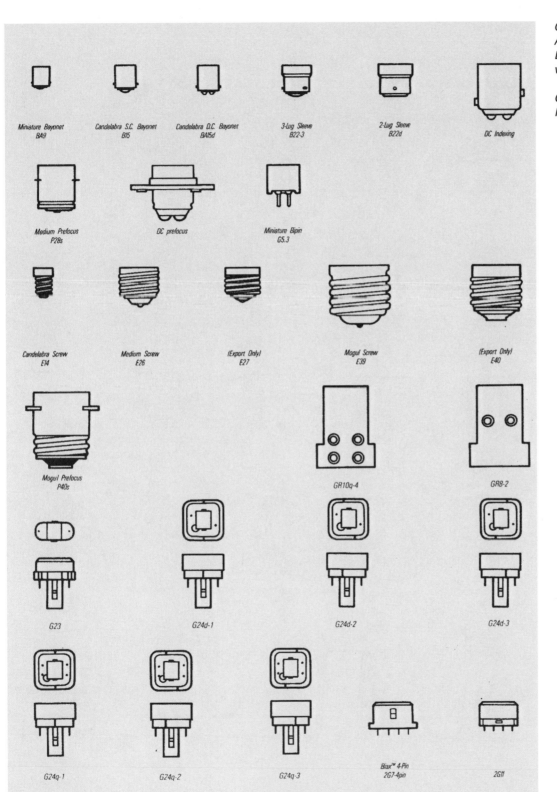

Chart 3.3
A Sampling of Lamp Bases
Lamp bases come in a wide variety of shapes and sizes.

Courtesy of G.E. Lighting, Nela Park

Chart 3.4
Here are even more
bases to study.

Courtesy of GE Lighting
Nela Park

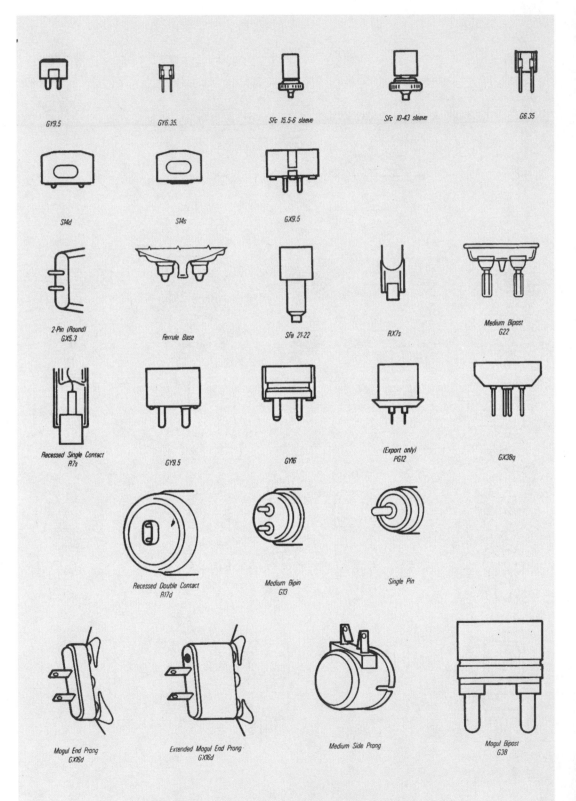

GY9.5

GY6.35.

SFc 15.5-6 sleeve

SFc 10-43 sleeve

G6.35

S14d

S14s

GX9.5

2-Pin (Round)
GX5.3

Ferrule Base

SFa 21-22

RX7s

Medium Bipost
G22

Recessed Single Contact
R7s

GY9.5

GY16

(Export only)
PG12

GX38q

Recessed Double Contact
R17d

Medium Bipin
G13

Single Pin

Mogul End Prong
GX16d

Extended Mogul End Prong
GX16d

Medium Side Prong

Mogul Bipost
G38

Chart 3.5
Filaments
There are a wide variety of filaments being used today. This is a good sampling for reference purposes.

Courtesy of G.E. Lighting, Nela Park

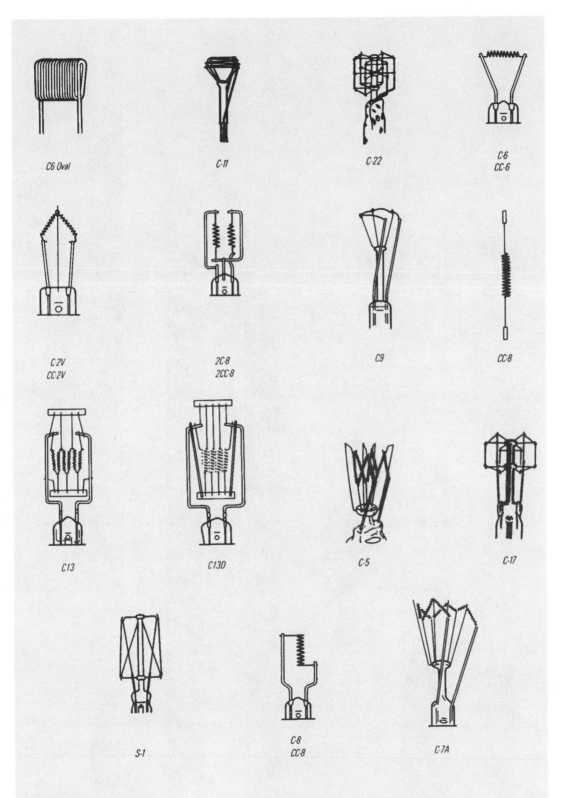

C6 Oval

C-11

C-22

C-6
CC-6

C-2V
CC-2V

2C-8
2CC-8

C9

CC-8

C13

C13D

C-5

C-17

S-1

C-8
CC-8

C-7A

Chart 3.6
Lamp Efficacy
This chart compares the various lamp categories and their respective efficacies. Mercury, metal halide and high Pressure sodium are H.I.D. (High Intensity Discharge) sources. Incandescent gives us the least bang for the buck.

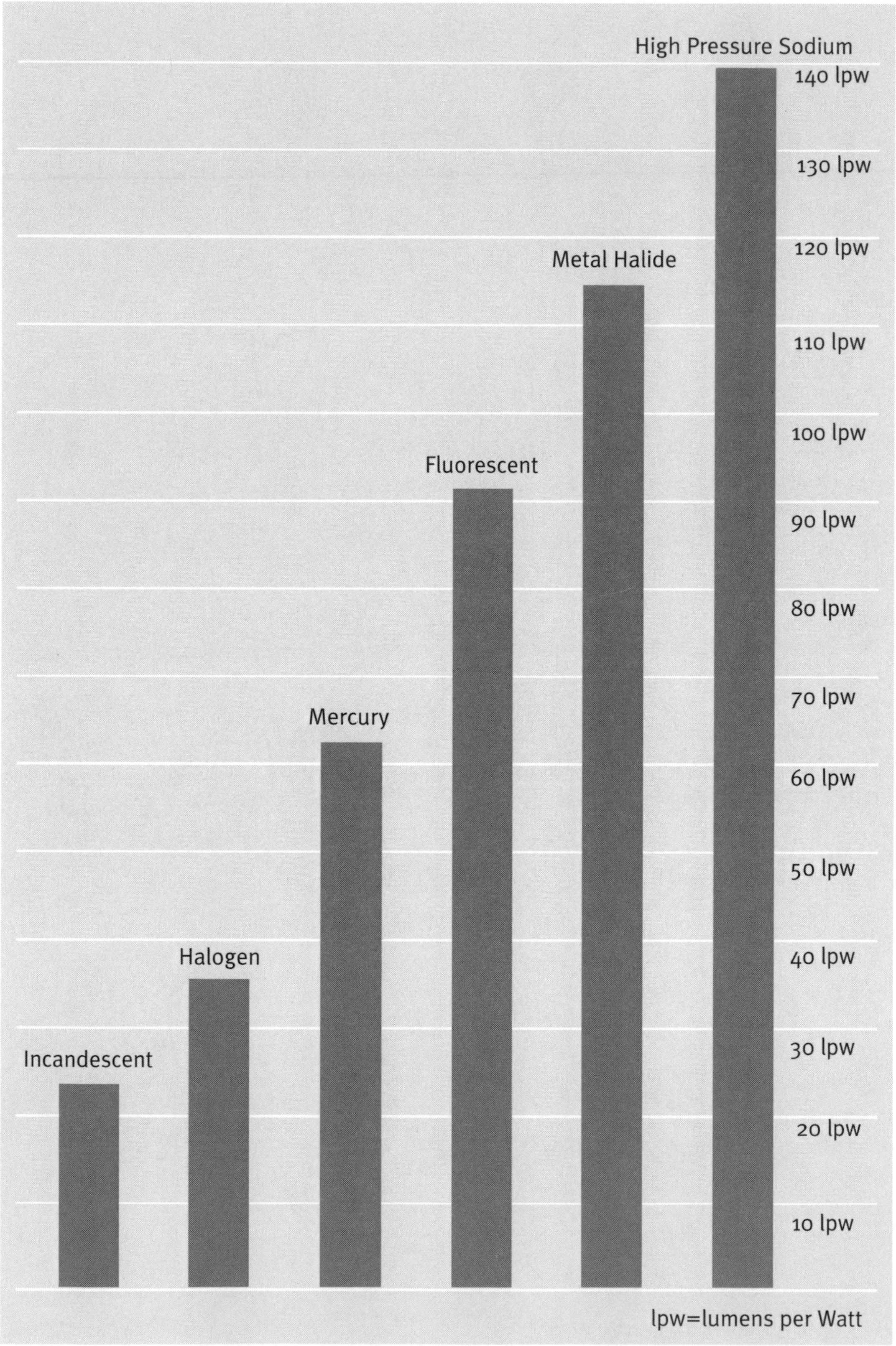

High Pressure Sodium

140 lpw

130 lpw

120 lpw

Metal Halide

110 lpw

100 lpw

Fluorescent

90 lpw

80 lpw

70 lpw

Mercury

60 lpw

50 lpw

Halogen

40 lpw

30 lpw

Incandescent

20 lpw

10 lpw

lpw=lumens per Watt

Chart 3.7
Efficiencies
Here is a comparison between a 40 watt household lamp and a 40 watt fluorescent lamp. The new lower wattage T-8 fluorescents are even more energy efficient.

4-foot fluorescent lamp (F40T12)

40 watts

Average rated lamp life: 20,000 hours

Over 200 Color options

Household A lamp (40A19IF)

40 watts

Average rated lamp life: 750 hours

Approximately 10 Color options

Standard incandescent lamps are:
1. A good point source with good optical control
2. Easily dimmed at a relatively low cost
3. Very versatile in regards to shapes and wattages

There are drawbacks too:
1. Short lamp life - 750 - 2000 hours
2. The least efficient of the three lamp categories (see Chart 3.6 and 3.7)

Yet we love it for that golden glow that takes us back to the time of candles, and maybe even further back to the days spent in our caves around a roaring fire.

Due to concerns over energy conservation and advancing technology, certain lamps have been replaced by newer, more efficient light sources. Don't brush off fluorescent sources.

It is important that you do not touch the glass envelope of a halogen lamp with your bare hands. The oil in your hands, transferred to the lamp when touched, could create a weak spot on the glass. This point of weakness in the envelope could cause the lamp to explode.

It is important that the fixtures you specify can adapt to the new lamps now on the market. R40's are now replaced by Par 30's. 40 watt cool white and warm white fluorescents are replaced by lower wattage lamps in T12 and T8 versions.

In 1995, certain lamps were outlawed by EPACT (the Energy Policy Act of 1992) including R40's and R52's as well as cool-white and warm-white fluorescents. These lamps have been replaced with more efficient light sources.

Halogen (also known as tungsten halogen or quartz)

Halogen is also an incandescent lamp and might be considered an advanced or improved incandescent lamp. "Just like standard incandescents but containing a halogen gas which recycles tungsten back onto the filament surface. The halogen gas allows the lamps to burn more intensely without sacrificing life" (Courtesy of G.E. Lighting).

There is a lot of misinformation about halogen that needs to be cleared up. It is often labeled a "white" source of light. That's a relative term. It is whiter than standard incandescent lamps by 200°K, but it's 2,000°K yellower than daylight. That's quite a large difference. Also, it is only whiter than standard incandescent when it is operating at full capacity. When dimmed, it becomes as yellow as any regular incandescent source. So, you need to treat halogen as basically a warm source of illumination in your design scheme.

Halogen Advantages
1. Halogen sources tend to be smaller in size than standard incandescent sources of comparable wattage.
2. Produce more light than standard incandescent sources of comparable wattage.
3. Have better optical control than most standard incandescent, fluorescent, or H.I.D. sources.
4. Come in a variety of shapes and sizes.

Halogen Disadvantages
1. Light yellows when dimmed as do all incandescent sources.
2. Dimming may shorten lamp life; lights should be turned up full for intervals to maximize lamp life.
3. The glass envelope should not be touched without gloves on.
4. Have to be shielded or enclosed in a glass envelope to protect the area around it from it's intense heat.

It is important that you do not touch the glass envelope with your bare hands. The oil in your hands, transferred to the lamp when touched, could create a weak spot on the glass. This point of weakness in the envelope could cause the lamp to explode. If you do touch the lamp, the surface can be cleaned with alcohol. Wearing gloves or using a clean cloth will prevent this problem. There are now double-envelope halogen lamps available that eliminate this handling precaution.

Here are some factors you should know about halogens:

Beam Spreads—With the right reflectors, halogen can produce a wide variety of beam spreads and a good punch of light.

Halogens are being made in almost all the shapes and sizes of incandescent lamps, along with a few that are unique, such as MR16's and MR11's. These particular lamps have been the hot ticket in the lighting world for the last 12 years. And their technology improves with each passing year. Their small size and beam control enable luminaire manufacturers to create a variety of new fixture styles.

MR16 lamps were originally made for slide projectors, so their catalog numbers often refer to the machinery for which they were designed. They have ANSI (American National Standards Institute) letters like EXN, EXT, and FJX. These are not as familiar to us as other lamps designations, which correlate to wattage, shape and size.

See Chart 3.8 for a list of MR16 and MR11 numbers and what wattage and beam spreads they represent.

Another group of lamps made originally for other uses are PAR 36's, available in standard incandescent and halogen varieties. They look like small automobile headlights. Not surprisingly, they were used as airplane landing lights, and fog lights on tractors. They, too, have ANSI designations that don't readily give us information about the lamp, while other PAR 36 lamps, produced later specifically for the lighting industry, do have the designations that are readily recognizable.

For example a 25PAR36 12V WFL means a 25 watt lamp with a parabolic, aluminized reflector that is 36 eighths of an inch in diameter. It operates at 12 volts with a beam spread that produces a wide flood of illumination. See Chart 3.9 for some of the more commonly used PAR 36 lamps.

Here are some straight answers to commonly-asked questions on halogen lamps:

What kind of wall dimmer can be used with low-voltage halogen lights?
It's best to use a dimmer that is specifically made to control low-voltage lighting. It will say on the dimmer box which is specifically for low voltage. Be aware that there are two types of low voltage transformers used with halogen lamps: electronic and magnetic. Make sure you choose a dimmer that is compatible with the system you select. Otherwise, the system may emit an audible hum.

Occasionally, darkening of a halogen lamp may occur. If this happens, simply turn on the lamp at 100% illumination for 10 minutes. The black residue (the result of tungsten evaporation) will disappear. Darkening of the lamp does not effect lamp life.

Note: All low voltage systems have some inherent hum. That hum comes from the transformer, the lamp, the dimmer, or any combination of the three. MR16 and MR11 lamps hum less than PAR36 lamps. If the transformers are installed in a remote location, then their hum is contained in the attic, basement, or garage space.

Chart 3.8
MR-11's and MR-16's

Multi-Mirror Reflector Lamps

MR11's	Wattage	Type	Beam
FTB	20-watt	MR11	10° narrow spot
FTC	20-watt	MR11	15° spot
FTD	20-watt	MR11	30° narrow flood
FTE	35-watt	MR11	8° narrow spot
FTF	35-watt	MR11	20° spot
FTH	35-watt	MR11	30° narrow flood

MR16's	Wattage	Type	Beam
EZX	20-watt	MR16	7° very narrow spot
ESX	20-watt	MR16	15° narrow spot
BAB	20-watt	MR16	40° flood
FRB	35-watt	MR16	12° narrow spot
FRA	35-watt	MR16	20° spot
FMW	35-watt	MR16	40° flood
EZY	42-watt	MR16	9° very narrow spot
EYS	42-watt	MR16	25° narrow flood
EXT	50-watt	MR16	15° narrow spot
EXZ	50-watt	MR16	25° narrow spot
EXK	50-watt	MR16	30° narrow flood
EXN	50-watt	MR16	40° flood
FNV	50-watt	MR16	55° wide flood
EYF	75-watt	MR16	15° narrow spot
EYJ	75-watt	MR16	25° narrow flood
EYC	75-watt	MR16	40° flood

Chart 3.9
Par 36 Lamps

PAR 36 shape lamps

LAMP	Wattage	Type	Beam
25PAR36/NSP	25-watt	PAR36	narrow spot
25PAR36/WFL	25-watt	PAR36	wide flood
25PAR36/VWFL	25-watt	PAR36	very wide flood
35PAR36/H/NSP8°	35-watt	halogen PAR36	narrow spot
35PAR36/H/VNSP5°	35-watt	halogen PAR36	very narrow spot
35PAR36/H/WFL8°	35-watt	halogen PAR36	wide flood
50PAR36/H/VNSP5°	50-watt	halogen PAR36	very narrow spot
50PAR36/H/NSP8°	50 watt	halogen PAR36	narrow spot
50PAR36/WFL8°	50-watt	halogen PAR36	wide flood
50PAR36/VNSP	50-watt	PAR36	very narrow spot
50PAR36/NSP	50-watt	PAR36	narrow spot
50PAR36/WFL	50-watt	PAR36	wide flood
50PAR36/WFL/4	50-watt	PAR36	wide flood
50PAR36/VWFL	50-watt	halogen PAR36	very wide flood

Numbered	Wattage	Type	Beam
4405	25-watt	PAR36	very narrow spot
4406	29.5-watt	PAR36	very narrow spot
4411	29.5-watt	PAR36	spot
4414	14.9-watt	PAR36	oblong spot
4415	29.9-watt	PAR36	oblong spot
4416	25-watt	PAR36	ovoid spot
7600	42-watt	halogen PAR36	ovoid flood
7606	50-watt	halogen PAR36	oblong flood
7610	50-watt	halogen PAR36	wide flood
7616	37.5-watt	halogen PAR36	ovoid spot

*NOTE: All PAR36 lamps listed above are 12 volts

Note: All low voltage systems have some inherent hum. That hum comes from the transformer, the lamp, the dimmer, or any combination of the three. MR16 and MR11 lamps hum less than PAR36 lamps. If the transformers are installed in a remote location, then their hum is contained in the attic, basement, or garage space.

A 50-watt halogen desk light with a high-low switch—What is the light output when on low?
The high-low switch cuts the light output approximately in half.

Do halogen bulbs last longer?
The average rated life for a halogen lamp is 2,000 - 2,500 hours. A standard "A" lamp is rated at 750 hours. Remember that average rated life means that, at the 2,000 - 2,500-hour mark, half the lamps will be burned out and half will still be working.

Do halogen bulbs use less energy?
Not really. The amount of money you'd spend to power a 50R20 lamp and a 50 watt MR16 is the same. The difference is that the MR16 lamp can produce a more concentrated beam of light, creating a better visual punch.

Do halogen bulbs use more electricity? Will you notice a difference in the electrical bill?
You can cut your energy bill by using smaller wattage halogen lamps to give a similar amount of light in luminaires that use higher wattage standard A-lamps and R-lamps.

Is halogen safe to use in a bathroom? Does moisture affect the bulb?
Yes, it is as safe as any other lamp used in bathroom applications.

Are halogen bulbs readily available? Where?
Lighting showrooms will carry a ready stock of halogen lamps. Hardware stores are beginning to carry a limited variety.

How can I determine the width of the beam from various sources?
Lamp and fixture manufacturers have graphs and charts to show beam patterns of different lamps in their fixtures. The narrowest beam spreads come from the PAR36 VNSP (very narrow spot) or the 4405 (pin spot). The next smallest beam spreads come from MR11 spots. Almost all lamps sources can provide a wide beam. It depends on how wide you want it and what type of lamp the luminaire you're using can accommodate. There is no set answer to this question (see Charts 3.8 and 3.9).

How far out from the wall should I locate the recessed cans?
Distance from the wall is determined by the height of the ceiling and the type of lamp and luminaire being used. Each luminaire manufacturer has charts to show distancing and spacing requirements for their luminaires. A longer shield on a fixture may restrict the beam spread.

Are there disadvantages to MR16 and MR11 lamps?
The dichroic reflectors project heat back inside the luminaire itself. The luminaire must be
designed to withstand that heat.

Also, the dichroic reflector projects a colored light out the back. Open or vented luminaires
will allow that color to be projected onto ceilings and wall.

These dichroic reflectors often vary the color of the light they project forward. Using one
company's lamps instead of mixing manufacturers will lessen this problem, but may not
eliminate it. MR16's and MR11's are available that have aluminized reflectors to help
combat the color fluctuations and backwash problem.

The circle of illumination from some of the tighter beam spreads may not be consistent.
Often, coronas of color and refractions can be seen along the perimeter of the beam of
light. The addition of one of the many spread lenses or diffusion filters can help soften
or eliminate this problem.

In the future, we will probably be seeing incandescent light being used primarily for accent
and decorative functions. Task and ambient lighting will fall on the shoulders of fluorescent and H.I.D. sources.

Flourescent
Fluorescent light is created by using electricity to energize a phosphor coating on the inside
of a glass envelope. Inside the envelope are droplets of mercury and inert gases such
as argon or krypton. At each end of the fluorescent tube are electrodes. When electricity flows between the electrodes. It creates an ultra-violet light. The ultra-violet light
causes the phosphor coating to glow or *fluoresce*, releasing the characteristic fluorescent light from the whole tube. The color temperature of the light will vary, depending
on the phosphors used.

Fluorescent lamps require a ballast to provide extra power to start the lamp and to control
the flow of electricity while it is operating. The early ballasts are partially responsible
for giving fluorescents a bad name—they hummed and caused the lamps to flicker.

The ballast is usually installed in the fixture housing but may be remoted in special situations where space is very limited. Some ballasts can handle more than one lamp.

There are two ways ballasts can be manufactured—pre-heat and rapid start. Pre-heat ballasts warm the electrodes to a glow stage activated by a starter switch (on the fixture
itself). These are used primarily in small fixtures such as undercabinet lights. Rapid
start ballasts have a circuit that continuously heats the electrodes, resulting in faster
illumination of the lamp. Almost all modern fixtures using 34-watt or higher lamps have
rapid-start ballasts. There are two variations of the rapid-start ballast.

Average Rated Lamp Life means that half-way through a test of a certain group of lamps, 50% of them are still working and 50% are burned out.

Magnetic - The first generation; unfortunately, they tend to hum and cause lamps to flicker.

Electronic or Solid State - They have very little if any hum and flickering is a thing of the past.

Before, a dimming ballast was very expensive and might flutter if dimmed beyond a certain point. Now, with the advent of new ballasts, many of the old problems have been resolved.

For dimming purposes, solid-state ballasts are the best available (and, of course, the more expensive). The dimmable version allows for 90% to full-range dimming (depends on the manufacturer), with no hum or buzzing, and you can dim unlike lamp lengths together. Before, you could only dim all four-footers or all three-footers together. The solid-state allows you to dim most lengths together. Chapter 6 will deal further with fluorescent dimming.

It is true that the fluorescent lamps we were exposed to as we were growing up were awful, but times have changed.

There were just two colors available in fluorescent for the longest time, cool white or warm white. Cool white gave you a greenish cast, while warm white gave an orange facsimile of incandescent. Their colors were obtained using different phosphor powders. These two choices were (and still are) just poor interpretations of the two ends of the Kelvin color temperature scale at least for bulbs commonly used. As noted before, both these lamps are no longer manufactured in 40-watt T12 version, for the U.S. Market. There are new improved fluorescents out there just waiting for you to embrace them and make them your own.

Now there are nearly two hundred colors available in fluorescent. Some are wonderful creamy peach hues, which are great for skin tone (but not the best for color rendering). You have so many colors to choose from that you can "paint" with light, from a very broad palette. These colors are obtained from a mix of phosphors.

Compact Fluorescents — Compact fluorescent lamps have opened up a whole range of uses that were not possible with larger-sized fluorescents. We are seeing them now in small diameter recessed luminaires, wall sconces, pendants, wall wash luminaires, and many more. This type of lamp is being improved more quickly than any other source on the market (with High Intensity Discharge sources running a close second).

The first compact units on the market had a noticeable hum, no rapid-start ballast, were undimmable, and had limited selection of color temperatures. Great strides have been made on all fronts. Today, there are rapid-start, quiet, dimmable compact fluorescent lamps in a variety of color temperatures. The dimming ballasts at this point are pricey, but they will come down in cost, just as the improved dimming of standard-sized fluorescent lamps did. See Chart 3.10 for the common fluorescent lamp shapes available.

Advantages of Flourescent Lighting:

Longer Lamp Life—A standard household "A"lamp has an average rated lamp life of 750 hours; the T-8 fluorescent is rated for 22,000 hours! The best that most standard reflector lamps can do is 2000 hours. A good MR16 gets 2500 - 3000 hours, while compact fluorescents are rated at 10,000 hours.

That's an enormous difference. It can be especially advantageous where the lamps are in a location where they're difficult to change when they burn out, not to mention the energy savings.

Lower Maintenance Time/Cost—Because fluorescent lamps last longer, they need to be replaced less often, saving time and money.

Average Rated Lamp Life means that half-way through a test of a certain group of lamps, 50% of them are still working and 50% are burned out.

More Lumen Output—Fluorescent lamps can produce three to five times more light for the same wattage as a standard "A"lamp.

For example, compare the light of a 40 watt household bulb and a four-foot fluorescent lamp. Both are 40 watts, yet you can visibly see how much more light you get from the fluorescent. (See chart 3.7)

Here again, you can see how using fluorescent lamps can produce significant energy savings because you don't need to use such high wattage lamps.

Cooler Source—Fluorescent lamps don't give off as much heat as incandescent sources. So not only will there be savings on air conditioning, but you won't have the problems of heat damage and fire danger as you may with high temperature sources like halogen. With color corrected phosphors, fluorescents make great light sources for closets. They can be installed closer to combustible material.

Color Variety—There are a huge number of color temperatures available in fluorescent lamps, while incandescent lamps are available in relatively few. So if you need a certain color temperature for a special space or setting, you'll be able to find it more easily using fluorescent lamps.

Dimming—Fluorescent lamps do not change in color temperature when dimmed as incandescent sources do, you don't have to worry that the whole color scheme will be altered when your client dims the lights.

Halfway through its life, a fluorescent lamp may produce 20% less light than when it was new.

Chart 3.10
Here are some of the
fluorescent lamp shapes
that are available.

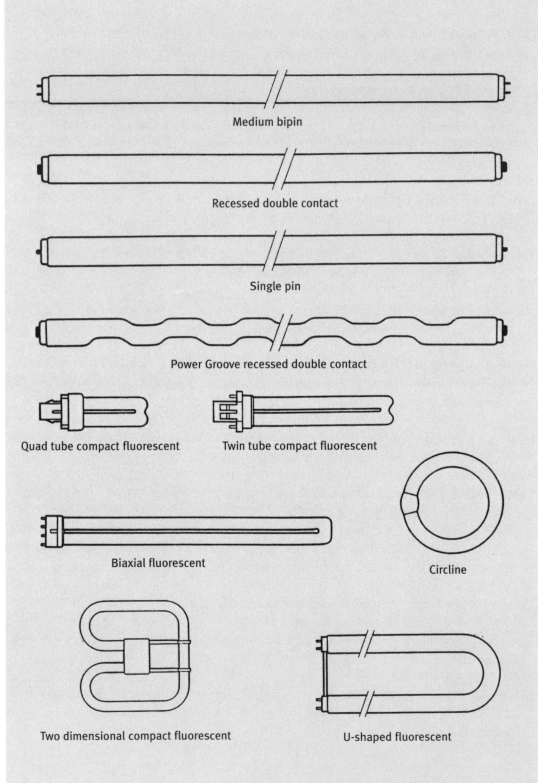

Medium bipin

Recessed double contact

Single pin

Power Groove recessed double contact

Quad tube compact fluorescent

Twin tube compact fluorescent

Biaxial fluorescent

Circline

Two dimensional compact fluorescent

U-shaped fluorescent

Disadvantages of Flourescent Lighting:

Fluorescent, like all lamps, are not perfect. They, too, have their drawbacks. Here is a listing of the ones you need to be aware of:

Hum—There are many fluorescent luminaires with inferior ballasts on the market. Be selective when you specify. Cheap luminaires hum and cheap lamps have reduced lamp lives.

Lamp Life and Lumen Output—Halfway through its life, a fluorescent may produce 20%
less light then when it was new. So, relamping at that point may be a good practice to
maximize the light output for the power consumed. Also, the ballast, which is integral
to fluorescent luminaires, will still use some power even if the lamp is removed. The
ballast must be disconnected if the energy savings are to be fully realized.

Relative inability to Accent—Fluorescent lamps are relatively large light sources. Through
the use of integral reflectors, manufacturers are able to achieve some success with flu
orescent luminaires, such as wall washers, for art and for uplighting trees. There are
even compact fluorescents inside reflector envelopes that mimic the directional capabilities of PAR and R lamps. But incandescent sources, such as MR16, PAR36's and
PAR20's, are still tops in the concentrated beam category.

Temperature Restrictions—A major disadvantage of fluorescents is their difficulty igniting
in very cold temperatures. They may start out very dim and take several minutes to
warm to full output. Below-freezing temperatures may keep them from igniting at all.

Prime Uses for Flourescent Sources:

Ambient Light—Fluorescent lamps can do a tremendous job of providing pleasing ambient
light. Their soft, even glow of illumination is well-suited to providing fill light. The right
color temperature and a solid-state dimming ballast can team up for a very usable, flexible ambient source of light.

Storage Areas—Fluorescent lamps are also extremely useful for storage areas. They are an
inexpensive way of producing a good amount of illumination. A color-corrected fluorescent luminaire mounted in the closet or laundry can provide accurate illumination
for color-matching and other such tasks.

High Intensity Discharge (H.I.D.)

This is the last of the three lamp categories and the one that holds the most mystery for
designers and architects.

High Intensity Discharge (H.I.D.) lamps are relatively easy sources to understand and may
end up being the perfect lamp selection for a specific aspect of an upcoming project.

There are cold-weather ballasts and well-sealed luminaires avail-able that should be used if the project is located in a region with below-freezing temperatures.

Caution: H.I.D. sources shift in color over their rated life.

See Chart 3.11 for H.I.D. lamp shapes.

The truth is that it's going to be a while before we see much H.I.D. being used for residential interiors. They are better suited for exterior lighting (both residential and commercial). They are large in size, require a ballast and have a limited number of wattages.

Technically speaking, inside the glass envelope of an H.I.D. lamp is a small cylinder (made of ceramic or quartz), called an "arc tube". It is filled with a blend of pressurized gasses. A ballast directs electricity through the tube and charges the gasses to produce light.

Each of the three kinds of H.I.D. lamps have their own special blend of gasses and produce different colored light.

Mercury Vapor has been around the longest. It produces a silvery blue/green light, terrible for skin tones but acceptable for lighting trees.

High Pressure Sodium is the most widely-used of the H.I.D. sources at present. Most of the streets in the United States are illuminated with high pressure sodium lamps. It emits a gold-orange light, which has poor color-rendering capabilities. Trees look dried out and dead, and people resemble the bottoms of copper cookware. Yet brick facades and sandstone walls (and even the Golden Gate Bridge) look great when illuminated by high pressure sodium.

Low Pressure Sodium (a discharge lamp) has even worse color gravity: gray-orange, which literally turns most colors to the same value. Of course this light source, which has the worst color, also happens to have the longest life and highest lumen output so it is unfortunately in wide use.

Metal Halide is the new kid on the block. It is the darling of the H.I.D. sources. It produces a light that is the whitest of the three types. It also comes in two very usable color temperatures, 3000K and 4000K.

A special advantage over the other H.I.D. sources is that metal halide lamps come in some very small sizes, allowing for very compact luminaires.

One of the main disadvantages of all H.I.D. lamps is their tendency to shift in color throughout their life. They don't all shift the same way. Metal halide will shift towards green or towards magenta, and this shift differs not only with each manufacturer but with each lamp as well. Mass relamping of the lamps halfway through their life will keep the color as constant as possible. Improved lamps, such as the Phillips Master Colorline™, have a shift that averages plus or minus 200K over the lamp's life.

Even as you are reading this, lamp manufactures are working hard to improve color quality, to reduce the size of mercury vapor and sodium sources, and to slow down the rate of

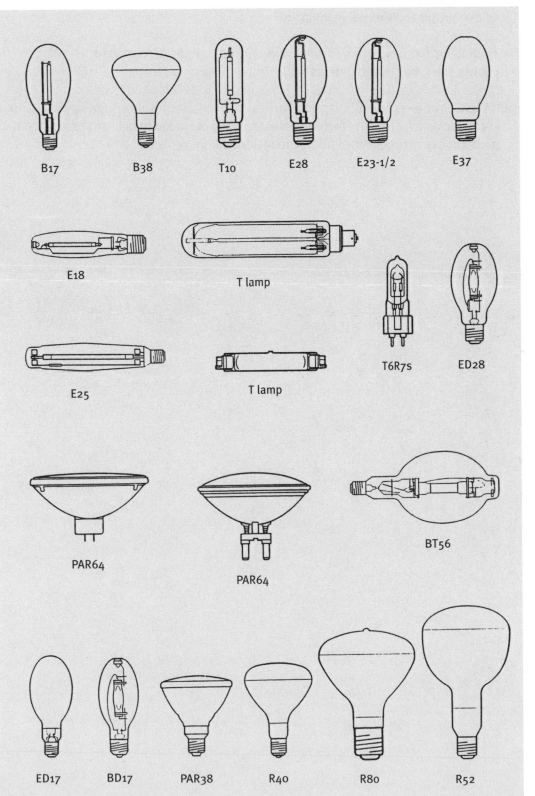

B17   B38   T10   E28   E23-1/2   E37

E18   T lamp

E25   T lamp   T6R7s   ED28

PAR64   PAR64   BT56

ED17   BD17   PAR38   R40   R80   R52

Chart 3.11
Here is a sampling of HID
(High Intensity Discharge)
lamp shapes.

color-shifting and extending lamp life.

Understanding the properties of the various lamps in all three categories will help you choose the correct luminaire, as well.

Good sources of more specific data on particular lamps are the lamp catalogues. The manufacturers offer a wealth of information, specially compiled to help designers and architects choose the right lamp for a particular need.

Chapter Four

Choosing the Correct Luminaires

Luminaires come in countless shapes and sizes. For purposes of this book, they will be divided into three categories: recessed, surface-mounted and portable.

Being familiar with the many types of luminaires that are available is as important as knowing what lamps should go into them.

Luminaires come in countless shapes and sizes. For purposes of this book, they will be divided into three categories: recessed, surface-mounted and portable (see Drawing 4.1 for some typical styles).

Portable Luminaires are the ones with which we are most familiar. They include reading lights, torchieres and uplights. If used correctly, they can be a quick fix to a good number of lighting problems. If they are misused, they can visually dominate a room, letting everything else fall into secondary importance (See Drawing 4.2). A portable luminaire has a cord and plug and can be easily moved from one location to another.

Table lamps are probably the most overused luminaires in residences. People want one

Drawing 4.1
Typical portable luminaires

Table lamp

Bouillotte lamp

Desktop pharmacy lamp

Banker's lamp

Stake light for plants

Uplight

*Drawing 4.2
Table lamps with translucent shades become the focal point of the room, forcing everything else to fall into secondary importance*

*Drawing 4.3
A pharmacy lamp provides excellent task lighting without drawing attention to itself*

*Drawing 4.4
A torchiere lamp is a fast, easy way of providing ambient light for a room*

luminaire to perform too many functions at once: decorative, accent, task light, and general illumination. This usually produces a sea of lampshades that draws your attention away from everything else.

Humans are naturally drawn to light; it is a part of our physiological makeup. For example: when you are driving down the freeway at night, you are confronted by a steady stream of oncoming headlights. They are glaring and uncomfortable to look at, but don't you look at every pair?

If you're designing a room, are linen lampshades the first thing you want people to see, or do you want them to see the fantastic art, the sumptuous colors you've chosen for the space, and the impeccable furniture selections you've made?

Table lamps can perform valuable functions if they are selected with care. A luminaire with a translucent shade (linen, silk, rice paper, etc.) works best as a decorative source of illumination. Using a low-wattage lamp (25 watts or less), they create little islands of light that draw people to seating areas and add a comforting human scale to a room.

If their primary function is to provide reading light, as mentioned in Chapter One, consider using a shade with an opaque liner and a perforated metal lid, so that the illumination is directed downwards onto the table and across one's lap. An opaque shade without the lid may cast a hard circle of light onto the ceiling, depending on the lamp being used.

Another consideration is a translucent white

opal diffuser on the bottom of the shade. This softens the light and shields the lamps from view.

Having these luminaires on dimmers or fitted with three-way lamps (a lamp that can be switched to 50, 100, or 150 watts) allows some light level flexibility. Hence they can be both decorative and task sources, depending on the need at the time.

A good alternative luminaire would be a *pharmacy-type* or metal tent-shaped shaded portable (in a floor or table style). Most of these luminaires have adjustable necks so that they can be easily repositioned for the particular height of the person using them. Selecting a luminaire with an opaque shade (normally metal) enables it to be positioned below eye level, providing shadow and glare-free illumination. The opaque shade also allows for significant light levels, without drawing attention to the luminaire itself (see Drawing 4.3).

Another portable luminaire that has been in demand is the *torchiere* (a floor lamp that projects light upwards — see Drawing 4.4). It is a quick, easy way of providing some ambient light for a given space. Select a luminaire that has an opaque shade. A translucent shade (such as glass) will draw too much attention to the torchiere itself.

Many torchieres are available with halogen sources and integral dimmers, making them energy-efficient with a wide range of light levels. They come in many styles to fit into most interiors from the very traditional to ultra-modern.

Uplights are another source of illumination for rooms that need some visual texture. An uplight, located behind a tall, leafy plant, will cast interesting shadows on the walls and ceiling, for a dramatic effect (see Drawing 4.5).

Sometimes, due to existing construction constraints (such as concrete ceilings) or budget constraints, uplights could be used to wash a painted screen with illumination instead

Drawing 4.5 (left)
An uplight behind a plant can help add texture and shadow to a room setting

Drawing 4.6 (right)
A series of uplights could do a pretty good job of washing a screen with illumination if lighting cannot come from the ceiling.

Drawing 4.7
A portable accent light on top of a bookcase could do an adequate job of illuminating a painting when budget or ceiling inaccessibility is a problem.

Drawing 4.8
A linear light source on top of a canopy bed could provide some subtle ambient light without immediately revealing the location of the luminaire

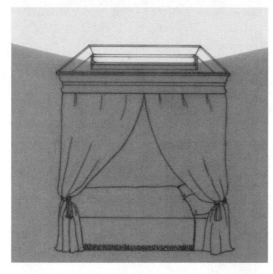

Drawing 4.9
Backlighting a translucent screen is another way of getting some ambient light into a room

of using recessed or track lighting (see Drawing 4.6).

Portable adjustable accent lights are another very flexible solution for highlighting objects without installing track or recessed luminaires. They can be tucked into bookcases or behind furniture to help add some focal points, giving added depth and dimension (see Drawing 4.7). Many track luminaire companies make weighted bases that accept their track heads. Ready-made units are also commonly available.

Where you place your portable luminaire can be a creative endeavor, in and of itself. For example, how about using one on top of a canopy bed, to provide some ambient light in a bedroom (see Drawing 4.8)?

Or, what about placing an uplight behind a translucent screen to add some visual interest to a dark corner and some soft fill light for the room (see Drawing 4.9)?

Portable picture luminaires are available in sizes to adequately light most pictures. Remember to look for picture luminaires that are adjustable front to back to accommodate different frame depths.

The battery operated units, while appealing because they do not require a cord and plug, have a very short lamp life.

Hanging fixtures that plug in and are portable are called swags. They utilize two hooks in the ceiling and the cord is woven through the chain. These were popular in the late 70 and early 80's, and may still be an option if a junction box is not feasible or placed incorrectly.

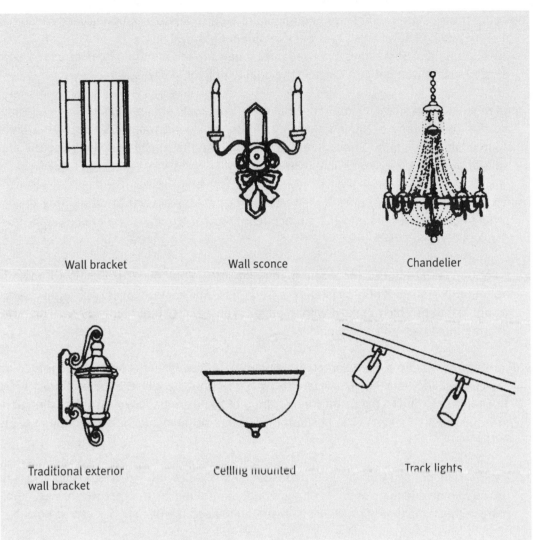

Drawing 4.10
Typical surface-mounted
luminaires

Wall bracket   Wall sconce   Chandelier

Traditional exterior   Ceiling mounted   Track lights
wall bracket

Please remember that portable luminaires are just one component of effective lighting design. Here are other choices for you to consider:

Surface-mounted luminaires are units that are hard-wired to a wall, ceiling, column, or even a tree.

Surface-mounted luminaires normally require a junction box, on which they are attached. These luminaires project out from (as opposed to being recessed into) the surface on which they are mounted. These include wall sconces, chandeliers, under-cabinet lights, and track lighting. (See Drawing 4.10 for some typical luminaire types).

In older homes the only source of illumination is often a surface-mounted fixture in the center of the room. Too often such a luminaire is called upon to perform all the lighting functions for that room. It can be an uncomfortable glaring light.

Please note that track lighting is almost exclusively a source of accent light — not ambient or task.

Replacing the existing ceiling-mounted luminaire with a pendant-hung indirect variety, as mentioned in Chapter One's section on ambient light, will fill the room with soft flattering illumination. Sometimes simply replacing one surface-mounted luminaire for another can make a world of difference in the quality of light in that space.

Chandeliers are also surface-mounted luminaires. The most common mistake is to let them be the sole source of illumination in an entry or dining room. Use additional light sources in the room for ambient and accent light so that the chandelier gives the illusion of providing the room's illumination.

Somewhere along the line, *track lighting* became the easy answer to all lighting design problems. Since it falls into the surface-mounted category, we can address that misconception here.

Track lighting is the solution for accent lighting in some situations, such as where there is not enough ceiling depth to install recessed units, and rental spaces, where the clients would like to be able to take it with them, or in an artist's studio-type space, where the lighting must be highly flexible.

Ordinarily, it cannot be a good source of ambient light, because it is normally mounted on the ceiling. The track luminaires are too close to produce any adequate indirect light. Track lighting, due to the usual configuration of the fixtures, is typically very directional. Care must be taken not to position the fixtures so the actual light beam hits people in the eyes.

If the track is suspended from a series of pendants or is wall-mounted, it is possible for it to provide some adequate ambient illumination. Should this be the case, use a track luminaire with a lamp that has a wide beam spread to keep the fill light as even as possible.

*Drawing 4.11
Mounting the track run on the side of the beam instead of on the bottom allows the beam itself to screen the luminaires from view when people enter the room*

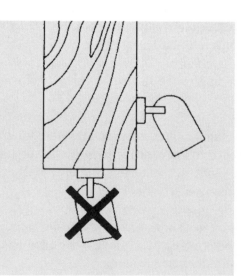

Track lighting is also a poor choice for task lighting, the main reason being that the beam of light produced by the track luminaire can't pass through your head, so it will cast a dark shadow onto your book or racing form. The track lighting fixtures are also visually intrusive. If there is another way to go (such as recessed lighting), consider taking that approach. If not, there is a technique to minimize the track's presence. If you have open beam construction (a flat or sloped ceiling where the beams are exposed), try to mount the track on the back side of the beams so that as people enter the room the beams act as a natural baffle for the track runs (See Drawing 4.11). Selecting a color for the track and the track fixtures that is as close to that of the mounting surface will minimize it's visibility. Then people will tend to look at what is being illuminated rather than inconspicuous fixtures on the ceiling.

Often, designers go with track because they fear that recessed lighting will be more expensive and actually penetrate the ceiling. This next section should allay some of those fears.

Remodel vs. New Construction

Now that people tend to stay in their present homes and upgrade them, instead of moving to larger homes or building new ones, designers are finding themselves involved with remodel projects more than ever. The more you do, the more you become accustomed to the problems that are inherent in this kind of work.

What designers may forget is how traumatizing remodeling can be for the homeowners themselves, especially if this is their first remodeling experience.

There is no getting around the sense of invasion that the clients feel and how such anomalies as plaster dust can literally touch every aspect of their lives.

The main thing they feel is loss of control. All of a sudden, they have less power over what is happening around them. This powerlessness, no matter how sub-conscious, can make them edgy, nervous and irritable.

Your job is to give them some sense of control over the process, such as: which doors the workers should use, what time they can begin work and what time they must stop, where they can park, which bathroom should be used, and whether workers should answer the telephone.

You may even go as far as suggesting a private job site phone/fax line and a rented portable toilet to save wear and tear on your clients. This is a minor investment that could help offset the problems that inevitably arise on every job.

You, as designers, know how much psychology is a part of your day to-day dealings with your clients. We are all fearful of the unknown and dislike unwelcome surprises. Informing them ahead of time what to expect and how they can involve themselves will allow the remodel process to proceed as smoothly as possible. And, face it, no project is problem-free.

On pages 47 and 48 is a list of expectations you can print and hand to your clients at the beginning of the project. Give it to them in writing, because we all have selective memory. Let them ask their questions up front; help them prepare for the onslaught:

What designers may forget is how traumatizing remodeling can be for home owners.

New Construction

Those of you lucky enough to be doing a project from the ground up can benefit from some pre-knowledge, as well.

First of all, incorporate the lighting design into the project right at the beginning. The earlier the better, preferably before the architect has finalized the plans.

Often, you receive the project after the drawings have been submitted to the planning department for approval and necessary permits. Normally the architect has drawn in a lighting plan, in order to make his submissions. In most cases these are specifications that are called out generically, such as "recessed luminaire", "surface-mounted luminaire", or "switch", without manufacturers' names or catalog numbers. They apply little to what you will be doing because you haven't designed the space. Changes in electrical plans can be made after permit approval without resubmitting the plans. The electrical inspector will simply refigure the permit fee based on the number of luminaires and outlets installed in the final design. It is important to keep local codes in mind when making these changes.

Suggested Handout:
The Realities of Remodeling for the Homeowner

At some point while living in your home, you looked around and realized it was time for some changes. It's almost inevitable, because most residences require updating as the homeowner's needs change.

Remodeling can strike fear into the heart of the most intrepid person. Remodeling has its inherent problems, but most difficulties can be greatly minimized with proper planning and by having the appropriate expectations.

Many of our clients have been through remodeling projects before, but for those of you that have not, we have put together this list of things to expect:

1. **References**—It is my job as your designer to recommend reputable trades people. If you have other recommendations or suggestions, I will be happy to contact them.

2. **Contracts** It is always a good idea to have a contract, spelling out what is expected from the contractor or trades person, what they will receive in return, and a completion date. This protects you legally, but also serves, more importantly, as a reference to remind the contractor what has been promised.

3. **Scheduling**—Since fixture delivery can be from 6 to 8 weeks, we need to select the fixtures and order them before we schedule the electrician. If delivery is delayed, the best option may be to schedule after the fixtures have arrived.

4. **Keep On Top of Things**—Whether you have contracted with me, as your designer, to oversee the remodeling or if you have hired a general contractor, it is important that you understand the timetables and check that the work is preceding as planned.

5. **Problems Will Come Up**—Don't let this scare you; it is the nature of remodel projects. If you haven't hired someone to work out the problems, you will need to deal with them yourself. Some people will have no difficulty with that, while others would prefer to have someone else oversee the project. Be prepared and you will be able to glide through your remodel with a minimum of hassle.

6. **Luminaires Made for Remodel**—There are many recessed luminaires that are made to fit into holes cut the same diameter as the luminaire itself. These round openings are made with a *hole saw*. In this way you can eliminate a lot of plastering and painting. Another special tool called a *right-angle drill* enables an electrician to feed wires from luminaire to luminaire above the ceiling line, so the ceiling can remain intact. The difficult part is to get from the luminaires to the switch locations. That usually involves opening up parts of the wall and ceiling in order to feed the wires through the fire blocking or other obstacles. An electrician specializing in remodel work will keep this opening-up to a minimum.

Please note that when necessary holes are made, a plasterer and painter will be needed to follow-up the electrician's work.

7. **Hidden Challenges**—No contractor can detect everything that's behind a wall or ceiling prior to starting a project. Sometimes there are factors that necessitate changing luminaire and switch locations, such as additional brace beams, shallow ceiling depths, or duct work. These occurrences are a normal part of a remodel project, and a certain amount of change should be expected.

8. **Plaster Dust**—There are few things more invasive in this world than plaster dust. Even if the rooms being remodeled are sealed off with plastic sheeting, the dust still penetrates other parts of the house. It is a problem that can be minimized, but never eliminated.

9. **Tradespeople in the House**—During the remodel process you will have a number of people working in your home. It is a good idea to establish house rules that will keep things comfortable. For example, you should designate which one door should be used, which one bathroom to use, and when the workers should answer the phone. Remember, they are people too, and sometimes a cup of coffee or a soda is greatly appreciated.

With these things in mind, I'm sure your remodeling project will go smoothly. Please feel free to contact me, your general contractor, or your electrician at anytime. Communication is important in any project.

Tools of the Trade

Lighting components and related installation tools for remodel projects have greatly improved to meet the needs of the remodel industry.

It used to be that in order for recessed luminaires to be installed into a ceiling, the contractor had to cut a 12" square opening in the sheet rock or plaster and lath, then use *bar hangers* (metal collars) to attach the housing to the joists. Next they would cut a series of holes or channels in the ceiling to run wire from unit to unit and down to the switch. Then a plasterer or dry wall contractor had to come in and patch and sand all the holes, followed by a painter.

Presently, many remodel projects can be installed with minimal patching or painting. Here are the components and tools one should be aware of:

Remodel Cans—There are housings (metal enclosures containing the sockets that go above the ceiling and hold the exposed trim) for recessed luminaires that are specifically made to be installed into existing ceilings. They can fit into a hole that is the same diameter as the housing itself. These housings are held in place by using metal clips to attach to the ceiling material, instead of the joists.

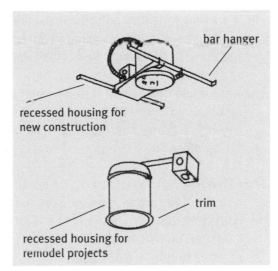

Drawing 4.12
Remodel cans allow for installation from below the ceiling, often without making a hole larger than the diameter of the luminaire itself.

The series of trims that fit into these housings have lips that are slightly wider than the housings so that they create a finished edge for the opening (See Drawing 4.12).

In any area where insulation might come in contact with the housing code requires the use of an IC-rated (insulated ceiling)housing. These housings are rated for direct contact with insulation and are large. They are not made in remodel versions.

Hole Saw—When remodel cans first came into the market, the contractors were still using a hand-held saw to make the rounded openings. This was a less-than-perfect method that wreaked havoc with plaster and lath ceilings, which tended to crack and crumble as the saw moved up and down.

Then a hole saw (originally designed to bore holes in doors in order to install hardware) was retooled to cut near-perfect holes in ceilings for remodel cans. It also does a great job of cutting holes in walls for junction boxes (See Drawing 4.13).

The hole saw is actually a special bit that attaches to a power drill. Good electricians will have various diameters as part of their equipment. Ask bidding electricians if they use them. Those who say they don't could cost your clients plenty in plastering and painting down the line.

Right-Angle Drill—Another great invention of the twentieth century is the right-angle drill. It allows the electrician to drill holes through joists above the ceiling line to wire from luminaire to luminaire with little or no opening of the ceiling beyond the holes cut for the remodel cans (See Drawing 4.14).

This drill comes with a series of 12" bits that can be linked together, allowing the electrician to drill up to four feet away from his starting position. This can be done from two directions, allowing the luminaires to be spaced up to eight feet apart without disturbing the ceiling between them.

There are longer flexible drill bits on the market, but the extra length could cause the end of the bit to end up poking through the new carpeting upstairs, or through the tile floor in the guest bath, one of those little surprises that should be avoided.

The right-angle drill should also be a part of a good electrician's tool collection, but it can be obtained from a tool rental place for a nominal fee of $10-$12 a day. This is the electricians responsibility. Again, ask the bidding electricians if they have used right-angle drills before hiring them.

Stud Finder—Calm down, it's not as exciting as you may think. The simple truth is that no one has X-ray vision. So how does the electrician know where to put a recessed luminaire if he doesn't know where the joists are located or how they are spaced. Not all joists run on perfect 16" centers. Sometimes there is additional blocking to support an upstairs fireplace, bathroom, or heart-shaped waterbed.

Drawing 4.13 (left)
A hole saw can make nearly perfect holes in walls and ceilings, even those made of plaster and lath

Drawing 4.14 (right)
A right-angle drill can make holes in joists above the ceiling line

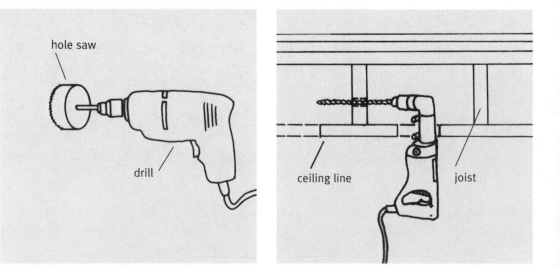

A stud finder is an electronic or magnetic device that senses the density changes in the ceiling (See Drawing 4.15). Using a chalk line or tape, the joist pattern can be laid out. If the ceiling has acoustical tile over dense plaster then locating joists will be difficult.

Wire Probe (Fish Tape)—Once the joists have been charted, the next step is to see if there are other obstructions that the stud finder didn't pick up (See Drawing 4.16).

What the electrician should do is drill a tiny hole and insert a flexible 1/8" diameter length of wire. The electrician feels around to see if there are obstructions and verifies that there is adequate depth for the particular recessed can specified. Sometimes ceilings change in depth, depending on what the floor above is supporting .

Understanding how these tools and components work will make you a better director of the work. You'll be amazed at the respect you get from an electrician when you use buzz words like "hole saw" and "right-angle drill". It's like asking your auto mechanic to check the timing because the engine is "dieseling", instead of saying "my car continues to run when I turn off the engine".

Furniture Layout—Your first step, as hard as it may be, is to come up with a furniture layout. Even if the exact design of the sofa or the finish of the dining room table hasn't been finalized, the location is important and greatly affects how your lighting is placed.

This is difficult, but without a semblance of a furniture plan, the lighting will neither complement nor be successfully integrated into your design.

For example, if you are floating a seating arrangement in the center of the living room and plan on table or reading lamps, you must know the location of the furniture and the type of floor covering in order to place the floor plugs.

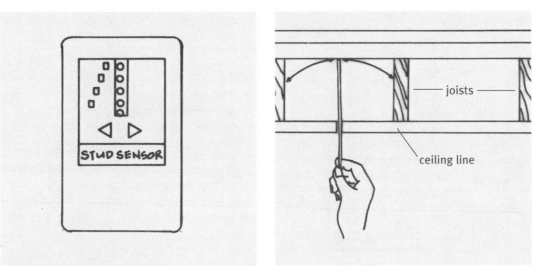

Drawing 4.15 (left)
A stud finder helps determine the joist spacing and obstructions by sensing changes in density above the ceiling line.

Drawing 4.16 (right)
A wire probe allows the electrician to check if any non-metal obstructions are in the way of a proposed recessed luminaire location

If a client cannot decide on one of your layouts, draw in the floor plugs with a note right on the plan that states "floor plug locations to be designated upon finalized layout of furniture. See owners/interior designer prior to installation." (See Drawing 4.17)

Another example would be the placement of wall sconces as the ambient light source in a dining room, Their location will affect the placement of art or tall furniture. Will they be placed on either side of the mirror over the sideboard, or will they flank the china hutch? Having a good idea of the size of the mirror or the furniture piece will determine the mounting location of the wall sconces.

Also, will the dining room table be centered in the room or centered in the space left after the sideboard is placed? This will set the location of the chandelier.

Another important factor is to study the elevations. Check which ceilings are sloped and how high they are. This will help determine which luminaires you use and where to place them. The higher a ceiling, the further out from the wall the recessed adjustable luminaires need to be placed in order to illuminate the center of the wall where the art will be displayed.

If the art is tall, the luminaires need to be closer to the walls in order to let some illumination reach the top of the piece. If you have a long tapestry that hangs from the ceiling line, you should use a wall-wash type luminaire that more evenly illuminates the full length of the weaving. Manufacturers' catalogues show spacing recommendations, based on ceiling heights. Get comfortable using these charts. They provide very useful information.

There are no hard and fast rules to luminaire placement. It is entirely dependent on what is in the space. The answer to the question "How many recessed lights do I need for a 12' x 18' room?" requires a series of questions:

Drawing 4.17
The location of floor plugs is determined by the furniture plan and the type of floor covering to be used. This is why it's important to have a furniture layout prior to implementing a lighting plan.

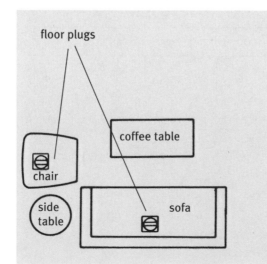

What is the room to be used for? Some functions require more light than others.

What is the age of the people living in the house? The older we get, the more we are adversely affected by glare.

How high are the ceilings? Tall ceilings allow wall sconces to be mounted higher, so art can be mounted below them.

Are the ceilings flat or sloped? If a ceiling slopes away from the wall you want to light, then a luminaire that compensates for that slope needs to be specified, such as a recessed can with an integral mirror.

What colors are the walls, ceiling and floor? Darker colors absorb more light, requiring higher levels of illumination.

Where will the art be? Determining the location of art pieces helps determine the number of accent lights needed.

Which way do the doors swing? Check the blueprints to see which way doors swing. This determines where your switches and dimmers should be located. Don't end up placing them behind the door: this is a common mistake that can be very irritating to the clients once they have moved in.

Door swings will affect your furniture placement as well. Double doors that open into the dining room will take up wall space (unless they are pocket doors). This will determine the visual center line of the wall on which you place furniture and art - which will, in turn, affect the location of the wall sconces.

These questions, plus others such as where furniture will be placed, need to be addressed before any quantitative lighting-related answers can be given.

As you well know, every design decision is based on a previous design aspect that has been addressed. Help your clients see how integral the furniture arrangement is to the lighting design. Color and style can be decided later: placement is the key.

Reflected Ceiling Plan—Study the reflected ceiling plan. It will show how the joists run, where the duct work will go for heating and ventilation, skylight locations, coffered ceilings, exposed beams, and all the other physical aspects that dictate where lighting can be placed. Specifying recessed adjustable luminaires allows some leeway in placement. You may want a tight spot of light in the middle of a coffee table. A recessed adjustable luminaire can do that without having to be dead-center over the table.

Also, people move their furniture and art around nowadays. A fixed recessed downlight has no flexibility, whereas recessed adjustable luminaires can be redirected to illuminate these components in their new locations.

Maintenance—Take into account accessibility of the luminaire locations you choose. If the owners can not easily reach a luminaire to change a burnt out lamp, they are less likely to do it.

For example, if you have a 20' ceiling in the entryway, don't put a recessed downlight in the center of the ceiling to illuminate the foyer table. Use a recessed adjustable luminaire, installed above the balcony railing on the second floor, to facilitate relamping (See Drawing 4.18). A long-life lamp is also advantageous.

> People move their furniture around nowadays. A fixed recessed downlight has no flexibility.

Low Voltage vs. Line Voltage

Many people are confused or intimidated by low-voltage lighting. There are a lot of misconceptions which should be cleared up in this section.

The National Electrical Code labels anything under 50 volts as low voltage. The most widely used systems are 12 volts (fixtures & transformers), while 6-volt and 24-volt systems are the next most common. By comparison, line voltage (also known as *house current*) is 110-120 volts. There are magnetic and electronic transformers. The magnetic have a lower failure rate but tend to be a little longer in size then the electronic. The electronic transformers are move compact and are used in small fixtures their size is a factor. A transformer is used to convert line voltage to low voltage. These transformers can be located within the luminaire (integral) or located elsewhere (remote). There are advantages and disadvantages to both.

Disadvantages to Low Voltage

Transformers and Voltage Drop—If you choose a remote transformer system, you will experience more voltage drop in the wire from the transformer to the housing, as the distance between them increases. The farther away the luminaires are from the transformer, the dimmer they become. If the distance between luminaires and the transformers is too great, it's best to use an additional transformer so that each luminaire is near a transformer. The magnitude of the voltage drop is directly proportional to the distance from the transformer and to the current, but this can be mitigated by increasing the size of the cable. If each luminaire has its own transformer, then there is no danger of voltage drop. However, this does increase the materials cost.

Also, remember that code requires that transformers be accessible. This is so that if one malfunctions, an electrician can easily reach the problem unit.

Limited Wattage—Most low-voltage lamps are available in 75 watts maximum wattage, with a few at 100 watts. This amount of wattage may not be enough to adequately illuminate larger objects, such as tall trees or long walls.

Drawing 4.18
It's better to locate a recessed adjustable luminaire over the balcony to facilitate relamping than in the center of a 20' high ceiling.

Hum—All transformers have some inherent noise: more transformers mean more accumulated noise. Remoting the transformer puts that noise in a more out of the way spot. For some people, the hum is barely perceptible; for others, it's a constant annoyance. Transformers are not the only source of noise. Particular lamps, such as many of the PAR 36's, as mentioned in Chapter 3, have their own audible sound. Also, some dimmers hum as well. Choosing components that are compatible will produce the quietest system. The main point is to inform your client first about potential hum, not after installation is complete.

Helpful Hints to Minimize Hum

1. Match the transformers and the dimmers. If you use luminaires that have electronic transformers, specify an electronic low-voltage dimmer. If you have magnetic transformers, use a low-voltage dimmer designed for magnetic loads.

2. Use a remote transformer system so that the noise is located in the basement, attic, garage or closet.

3. Specify luminaires that use MR16 or MR11 lamps, to avoid the vibration of PAR 36 lamps. The MR lamps tend to be much quieter. Lamp filaments can be a component in the amount of noise produced by a lamp.

4. In using recessed luminaires, make sure to specify a luminaire that has the transformer separate from the housing or with some type of flexible mounting between the housing and the transformer. Otherwise, the vibrations of the transformer may create a resonance in the housing which amplifies the sound.

5. In a room that has hard surfaces, add sound-dampening materials, such as curtains, plants, carpeting, or acoustical tile.

Advantages to Low Voltage

Size—Low voltage lamps can come in very compact sizes. These tiny lamps allow for fairly small luminaires. These luminaires are easier to tuck out of the way or allow for minimal openings in ceilings. Landscape, track, and recessed luminaires include examples of small fixtures using low voltage lamps for size advantage.

Beam Spread—Low-voltage lamps come in a great variety of beam spreads, from very tight spots of light to wide floods of illumination. Matching a beam spread to the proportions of a particular painting, table top or plant will make it stand out dramatically, and help add dimensionality to a room.

Accessories—There are many optional accessories for basic luminaires. Here are some that can help a particular luminaire do the best possible job of illumination:

1. *Louvers* are designed to help cut glare at its source by shielding the light. There are three basic types available: concentric ring, egg crate, and honeycomb (See Drawing 4.19). Check with the luminaire manufacturer for the appropriate louver holder to specify, some louvers are held in place by pressure.

Louvers are designed to help cut glare at its source by shielding the light.

Rule of Thumb:
Try to get the
louver as close
to the face of the
lamp as possible.
This will allow
for the greatest
amount of
illumination. The
further away the
louver is, the
more illumination
you would be
blocking.

The concentric ring, which looks like a gun-sight, was the first one developed and is, for the most part, useless. The egg crate is an improvement over the concentric ring, but is not the best. The ultimate is the honeycomb louver. They can be made for tiny MR11 lamps or huge H.I.D. luminaires.

Rule of Thumb: Try to get the louver as close to the face of the lamp as possible. This will allow for the greatest amount of illumination. The further away the louver is, the more illumination you would be blocking.

2. *Filters*—There are color filters available for a majority of the luminaires on the market today. One of the most popular is the color-correcting "daylight blue" filter mentioned in Chapter 2. This light blue filter helps minimize the amber hue of incandescent light and to produce a whiter light. This white light renders art and plantings in a more natural color, truer to a daylight or moonlight environment. Remember though, that people do not look good under a whiter light, so use this color-corrected filter for objects only, not your clients or their guests.

There are also peach-colored filters, sometimes known as cosmetic filters, that are very complimentary to skin tones.

Many other filter colors are available as well. These tend to make a strong statement and should be used judiciously. They also tend to cut down on the amount of light a luminaire can produce, depending on intensity of the color. Blue, green, red, and amber are industry standards.

There are also MR16 lamps on the market that are coated to produce specific colors. The advantage is that you don't have to buy a separate filter. The disadvantage is you are committed to that one color. A clear lamp can be adapted to project many colors simply by changing the filter. Try to select dichroic colored lenses; they hold up best and give

Drawing 4.19
The best of the three louver types available is the honeycomb

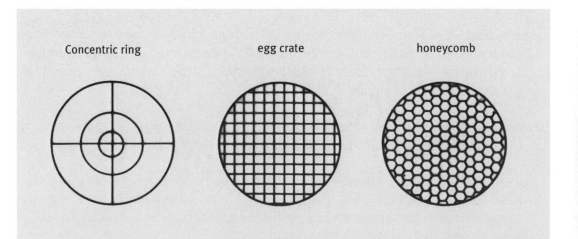

Concentric ring   egg crate   honeycomb

truer colors. Be aware that you may get a rainbow effect at the edges when a MR flood lamp is used.

3. *Lenses*—As opposed to filters, lenses are not colored. Instead, they are etched, formed or sandblasted in various ways to alter the beam spread of a given lamp. Here are some examples:

Spread Lens—These are pieces of glass that are sandblasted or patterned to widen the overall circle of light produced by a particular lamp. They also help soften the edges of a spread of light. Use these to avoid a sharp cutoff of light.

Linear Spread Lens—This type of lens stretches a beam spread either vertically or horizontally, depending on how the lens is positioned in front of the lamp. These work particularly well for a rectangular painting.

Fresnel Lens—This is a holdover from theatrical lighting. It also helps diffuse the light patterns, as does a spread lens. Fresnel lenses are used primarily with recessed luminaires and some track systems.

Suggestions

Now you have a good base of information on which to make your luminaire selections. Try to build your luminaire catalog library and keep it updated. Lighting showrooms are your best source for new catalogs, seeing new fixtures, and many conduct seminars on new lighting products.

Organizations such as the I.A.L.D. (International Association of Lighting Designers), the I.E.S. (Illuminating Engineering Society), and the A.L.A. (American Lighting Association) give classes and seminars on lighting-related topics. Attending these functions will help broaden your understanding of lighting design.

Industry-related magazines like *LD&A (Lighting Design and Application)*, *Architectural Record Lighting* and *Lighting Dimensions*; are excellent sources of techniques and aesthetic information.

Chapter Five

DAYLIGHTING—INTEGRATION OF NATURAL LIGHT

Natural light is a wonderful source of illumination. Used correctly it can transform a dark dreary house into a bright, inviting sanctuary.

Daylighting is an all-natural source of illumination that is available to everyone. Making optimal use of daylight can certainly brighten and open up a room or area. Designers should explore the options of daylight. Here are some considerations:

Windows are the first option when thinking about getting natural light into a room. Windows do a fine job of providing daylight for rooms that run along the perimeter of the house. There is, however, the problem of sun damage (permanent fading and dis-integration of natural fibers and woods). The sun's ultra-violet (U.V.) rays eventually harm the furnishings and surfaces within the home. All sources of light damage photo-sensitive materials, but U.V. light does the most damage.

Window treatments, such as shades, drapes, sheers and mini-blinds, do a good job of dif-fusing the light and slowing down the process of ultra-violet damage. Shutters and black-out shades do a better job of blocking the light that is not needed, such as when the homeowners are on vacation. Quantitative information on recommended maximum exposure to U.V. light is available in the IES (Illuminating Engineering Society) Handbook, which is a great source of technical information on industry standards.

Some companies offer an ultra-violet film as a ready-made product that's applied directly to the window or sandwiched in-between two panes of glass. These companies claim to cut 99% of the sun's harmful rays. What they don't tell you is that any standard glass or Plexiglas® will filter out 98% of the ultra-violet light. This is just an additional 1% over what standard glass already accomplishes. If it would normally take 5 years for the sun to take it's toll, the extra 1% of U.V. protection in the film would double the life of your furniture, rugs, etc. That's a pretty good deal. Remember, though, it's not only day-light, but all sources of light that contributes to the damage.

You can't eliminate the damage unless you use "black-out" shades all day, and don't turn on any lights which could be somewhat depressing for your clients. But at least you can slow the process down. There are manufacturers of motorized shades that can help control the natural light to the amount needed. Different types of shade materials can be scrolled together to offer a variety of options; there is even one that allows the shade to come up from the bottom for privacy and create a "clerestory" (a long narrow window extending along the top part of a wall) allowing natural light to come in without loss of privacy (See Drawing 5.1).

Windows are also wonderful visual portals to exterior spaces, making the inside rooms seem larger.

Making good use of windows and exterior lighting can also work wonders in making the exterior landscaping and view part of the interior space, especially at night. How often have we entered a wonderfully designed interior only to find that the windows are all *black holes* at night? Exterior lighting will help minimize those reflective surfaces, and enlarge the interior at night by visually extending your view out into the landscape. Exterior lighting techniques will be addressed in Chapter 15.

Doors, especially interior doors, do not have to be solid. French doors (they come in many styles), fitted with clear or frosted glass, allow light to travel into the room beyond and or though the space, so the natural light is shared.

Skylights are a great source of natural light, but should be used with a few precautions. They come in three standard varieties: clear, bronze, or white.

The clear variety can result in the most severe sun damage. All three varieties come standard with an ultra-violet inhibitor, or it can be ordered as an option. Clear skylights, although the most popular, cause the most problems:

1. The daylight coming through the skylight will be projected in the shape of the skylight. Your clients will be confronted by an intense square, circle or rectangle of light. The pattern of the sun will eventually be traced on the surfaces of the furniture, floors and wall coverings.

2. The light coming through will be a hard, harsh light that creates unattractive shadows on people's faces.

3. The clear skylights show dust and dirt immediately, so maintenance is a constant problem.

If that type of skylight already exists, one

Drawing 5.1
A shade that raises and lowers from the bottom allows light to come in without loss of privacy

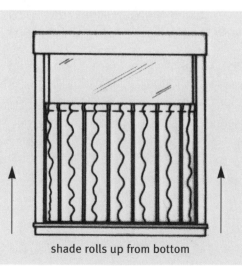

shade rolls up from bottom

solution is to get a pleated or Roman-style shade to help diffuse the light. This is especially good for operable skylights, so the shade can be pulled back to allow for ventilation when needed (See Drawing 5.2).

If the skylight is non-operable, you can install a white acrylic or glass diffuser within the light well to help spread out the sun's illumination. If there is enough room in the light well above the shade or diffuser, luminaires can be installed to keep the skylight from becoming a negative space at night (See Drawing 5.3).

The same considerations also apply to existing bronze tinted skylights. They cut down tremendously on the amount of light coming through. The main concern, though, is not reducing the amount of light but of diffusing the light for a more user-friendly illumination.

In new construction or a remodel project that will include installing skylights, you should consider using the white (also known as opal or frosted) skylight, because:

1. It distributes the light more evenly in a room.
2. It doesn't project a light pattern.
3. It doesn't become a black hole at night.
4. It minimizes maintenence.

Glass Block (also known as glass brick) is a building material that was popular in the 1930's to the 1950's and then fell out of popularity. The post-modern influence of the 1980's brought about a resurgence of its use. Succinctly put, glass block turns walls into windows. They come in clear, patterned, curved, frosted, colored, solid, or hollow varieties. This allows for varying degrees of light transference and varying degrees of privacy. Rooms with little or no natural light could benefit greatly from the installation of a glass block wall which allows them to borrow light from adjoining rooms with windows or skylights.

> In a project that will include skylights, consider using the white variety instead of clear or bronze.

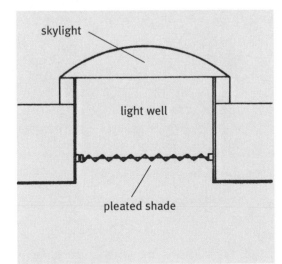

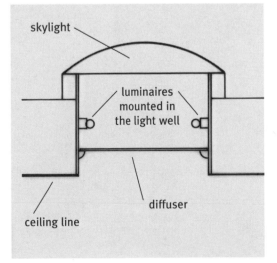

Drawing 5.2 (left)
A pleated or Roman-type shade fitted into the light well will help diffuse the daylight flooding into the room

Drawing 5.3 (right)
Lighting in the light well above the diffuser keeps the skylight from becoming a "black hole" at night

Don't try to light a glass block wall. It is a translucent product, so light passes through it. At best, you would highlight the grout in between the blocks, which is not the most visually ornate aspect. If the blocks are frosted, then the wall can be directly illuminated.

Note: Don't try to light a glass block wall. It is a translucent product, so light passes through it. At best, you would highlight the grout in between the blocks, which is not the most visually ornate aspect. If the blocks are frosted, then the wall can be directly illuminated. If the block is transparent, then the trick is to light the opposite wall. What you end up seeing is the illuminated wall through the glass block. This creates the illusion that the block itself is illuminated (See Drawing 5.4).

Reflectance: Don't forget that light colors reflect more light than dark colors. A Navajo white room will make better use of the natural light than one painted a hunter green.

Mirrors are another tool to help make the most of natural light. A room with windows on one side can greatly increase the perceived amount of daylight by installing mirrors on the opposite wall. Mirrors can also greatly increase the apparent depth of an area, by giving the illusion that the interior extends much further than it actually does. Cramped areas can become brighter and more open.

Areas such as kitchens and family rooms that are used frequently during the day are prime candidates for maximizing natural light.

Closets and laundry rooms that have some source of natural light make it easier to color-match blouses to skirts, ties to suits, and high heels to slacks. Be sure to take precautions to block the natural light when it's not required. Silk and wool are quickly damaged by light.

Natural light is a wonderful source of illumination. Used correctly it can transform a dark dreary house into a bright inviting sanctuary.

Drawing 5.4
Glass block appears to be illuminated if the wall behind it is washed with light

glass black

recessed wall wash luminaire

back wall

Chapter Six

CONTROLS—DIMMING, SWITCHING, TIMERS, PHOTOSENSORS, MOTION DETECTORS

Every room is capable of being viewed and used in multiple ways. Clients may want subtle mood lighting for entertaining, yet also want the ability to increase light levels for cleaning. Multiple scene controllers can do this at the push of a button.

Once you have formulated a lighting design, the next step is to decide how to regulate the system or control the lighting design to achieve the desired effect. There are many options: from simple switches to preset multiple-zone/multiple-scene controllers to turning on lights from a car phone. Every room is capable of being viewed and used in numerous ways. A multiple scene controller allows the same lights to be illuminated at various levels by a single control unit (see drawing 6.8). Clients may want subtle mood lighting for entertaining, yet also want the ability to increase light levels for cleaning. Multiple scene controllers can do this at the push of a button. The options may be almost limitless, but they need to be matched to the wants and capabilities of your clients. Designing a dimming system that is too complicated leaves the clients intimidated and ultimately frustrated. Try to meet their needs with a simple-to-operate control system. Also remember that they are the ones that will be living in the house, so the controls must be logically placed to match their traffic patterns.

A common mistake is to locate a majority of the controls at the front door. In reality, most people don't enter their houses through the front door; they park in the garage and enter through the garage. This spot should be your starting point. On the other hand, people who live in condominiums do enter by way of the front door. Each project is different, as are the needs of your various clients. Give them options and let them tell you what they want.

Switch Control

Today, people can put dimmers on most everything, but there are some instances where simple switching is the best choice or, more importantly, a code requirement. Low activity rooms, such as closets, storage rooms, attics and pantries, don't require dimmers. These rooms don't need varying light levels. If your clients have children, or are on the forgetful side, you might recommend these alternatives to the simple toggle switches we grew up with:

1. **Momentary Contact Switches** (door jam switches) are devices that turn on the light when the door is opened. It's the same way the light in your refrigerator operates. When the door is shut, the light automatically goes out. This only works if the clients, their children and other members of their extended family remember to shut the door completely. Bi-fold doors don't work well with momentary contact switches because they need to be fully closed to make a good connection.

2. **Motion Sensors** turn the lights on as someone enters the room and keeps them on as long as there is movement or for a certain length of time, as preset into the sensor. Some companies make motion sensors with a manual override, for clients who plan on spending a great deal of time in a particular room without moving around. For example, Let's say there is a safe in the master closet and your clients occasionally review important papers, or sort through jewelry. In this instance, an override on the motion sensor would be a good idea.

3. **Panic Switch** is a switch normally located in the master bedroom that turns on exterior perimeter lights. Nobody wants to run downstairs to the door when there is a suspicious noise outside at night. Most people don't even want to get out of bed. This will also allow you to turn off the outside or entry lights if they were left on inadvertently.

4. **Half-Switched Receptacles,** also known as *half-hot plugs*, can be a terrific labor-saving device in houses that use a lot of portable luminaires. In a half-switched receptacle, one of the two outlets is activated by a wall switch and the other outlet is continuously live.

This allows the clients to plug their table lamps, torchieres, reading lights, china hutch lighting or uplights into the switched receptacles. So, instead of going from luminaire to luminaire to turn them all on or off, they would all come on at the flick of one switch (See Drawing 6.1).

Drawing 6.1 (left)
Half-switched receptacles help combine a number of portable luminaires into one switching group

Drawing 6.2 (right)
A socket dimmer is a very easy way to make an existing table lamp dimmable

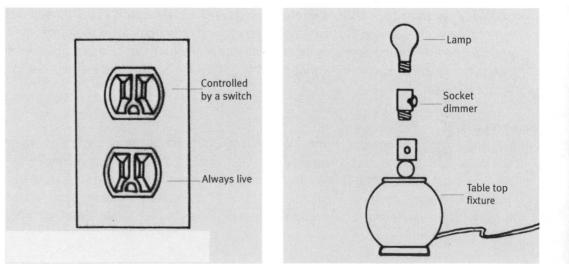

Controlled by a switch

Always live

Lamp

Socket dimmer

Table top fixture

Those items that they don't want to be turned on and off, such as clocks, televisions or stereo equipment, would be plugged into the non-switched receptacles.

Caution - do not put the half-switched receptacles on a dimmer, because if someone plugged a television or vacuum cleaner into a dimmed receptacle, it could damage the appliance or be a possible fire hazard.

If your clients want to dim their table lamps, have them purchase ones with integral dimmers. If they already have the table lamps, they can be retrofitted with dimmers or screw-in type socket dimmers (see Drawing 6.2) or cord dimmers (see Drawing 6.3).

5. **Sound-Actuated Switches** are devices that turn on lights when a noise is made. It's like the units you see advertised on television. They are a low-end solution to half-switched receptacles, with one inherent problem: If your clients or their guests make noises that are as loud as clapping, they may find the lights turning on and off at an inopportune moment.

6. **Three-way switching** is a term that confuses many homeowners and designers alike. The natural impression is that a light or group of lights can be turned on from three locations, which is incorrect. The correct answer is two locations. Why? Because a single location switch has two terminals and a three-way switch has three terminals; it does *not* refer to the number of control spots. The actual luminaire counts as one leg in the switching chain.

Example: If you have a hallway with a switch at both ends. One of these switches can be replaced with a three-way dimmer to allow for a variety of light levels. The three-way switch at one end of the hall will turn on the lights at whatever level the dimmer at the other end of the hall has been set (see Drawing 6.4). In simple three-way setups, you can't put a dimmer at both ends. More sophisticated (and more expensive) systems would allow dimming at both ends.

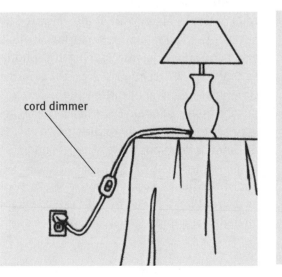

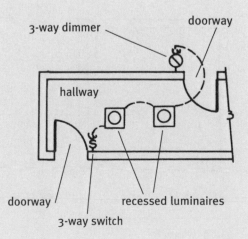

Drawing 6.3 (left)
A cord dimmer is a quick way of obtaining a full range of light levels from an existing luminaire

Drawing 6.4 (right)
A three-way set-up means that the lights are controlled from two locations (not three)

A four-way switching system allows three or more locations to turn a light or groups of lights on and off (see Drawing 6.5)

Timers

Timers are another option for turning lights on and off. Clients who travel or are cautious about security will want their home to look occupied at night. Lights controlled by timers that go on or off, at staggered times, when no one is home. Here are your options:

1. **Plug-In Timers** are readily available at hardware stores and home improvement centers. You plug them into a live receptacle, then plug a portable luminaire such as a table lamp into it. Then the device is manually set to switch on and off at specific times. Using a few in the living room, set to come on at dusk and go off around 11:00 P.M., then having one go on in the bedroom at 11:05 P.M. and staying on until midnight, gives the impression of people moving from a gathering spot to bed in a natural progression.

2. **24-Hour Programmable Timers** simply turn specific lights on and off at the same time each day. They are available in hard-wired (permanently installed) varieties, as well as plug-in. They release a regulating switch located inside a control box placed in the home, located in the garage. This unit along with the plug-in version turn the lights on and off at the same time everyday. It is more evident that no one is home and timers are being used.

3. **24-Hour/7-Day Programmable Timers** allow for different settings each day and give the option to skip days, for a look that seems to be controlled by a person instead of a device. For example, on Monday the lights come on at 7:15 P.M. and go off at 11:00 P.M. On Tuesday, the lights come on at 6:45 P.M. and go off at 12:10 A.M. Each day can be a little different. This is great for weekend residences and for clients who travel frequently.

Drawing 6.5
A four-way set-up indicates that there are three or more switches that can turn on a specific light or group of lights

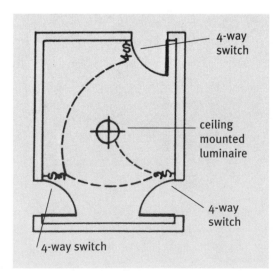

4. **Photo-Sensors** (also known as photo-cells or photo-electric cells) are activated by any light. The luminaires come on at dusk and turn off at dawn. Cleaning the photo-sensor itself regularly is a priority. Otherwise, as it gets dirtier, the cell reads dusk coming sooner. Eventually the lights may stay on all the time. Be sure to locate the sensor in a spot that is free of shade, so that it can detect true dawn and dusk, yet not in the direct path or another light, such as a street light. The unit will "see" daylight all night long and never come on.

Dimmers

Dimmers are a form of switching that allows variable adjustment of light levels. (See Drawing 6.6) Dimmers come in many varieties, such as rotary, toggle, glider, touch, low voltage and line voltage.

Here are some points that will help you and your clients make an informed decision:

1. Choose a dimmer that is specifically made for the luminaire type that you are dimming. For example, use a low-voltage dimmer when dimming a low-voltage system. If the low-voltage luminaire uses an electronic (solid-state) transformer, the dimmer must be electronic as well, or there will be a audible humming noise and possible damage to the dimmer or fixture. If the low-voltage luminaires use magnetic transformers, then the dimmer must be designed for an inductive load.

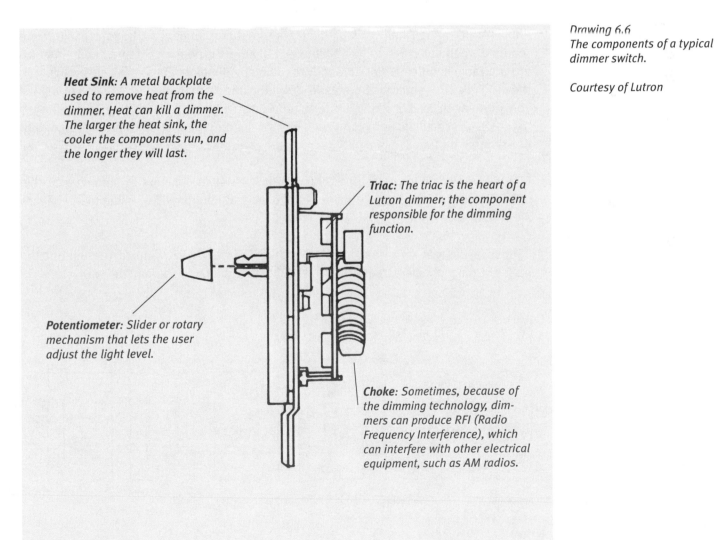

*Drawing 6.6
The components of a typical dimmer switch.*

Courtesy of Lutron

Heat Sink: *A metal backplate used to remove heat from the dimmer. Heat can kill a dimmer. The larger the heat sink, the cooler the components run, and the longer they will last.*

Triac: *The triac is the heart of a Lutron dimmer; the component responsible for the dimming function.*

Potentiometer: *Slider or rotary mechanism that lets the user adjust the light level.*

Choke: *Sometimes, because of the dimming technology, dimmers can produce RFI (Radio Frequency Interference), which can interfere with other electrical equipment, such as AM radios.*

The mounting of two or more controls side by side within one enclosure is called ganging. De-rating is the reduction of the maximum capacity (hood) a dimmer can reliably handle when the side sections (fins) are removed.

Note: Some low voltage electronic dimmers have only a 300-watt capacity as compared to a standard 600-watt low-voltage or line-voltage dimmer. So care should be taken in the number of total watts each dimmer can handle.

2. Be aware of the maximum wattage per the dimmer. A normal dimmer is rated for 600 watts. Putting 600 watts on that dimmer requires it to work its hardest. Most manufacturers will provide guidelines for load limits. For example, a Lutron low-voltage dimmer, such as the "Skylark" SLV600P, can optimally handle 450 watts worth of light luminaires.

The mounting of two or more controls side by side within one enclosure is called "ganging." Ganging dimmers together decreases their load capacities. De-rating is the reduction of the maximum capacity (hood) a dimmer can reliably handle when the side sections (fins) are removed.

3. Allow for de-rating when banking dimmers together. Dimmers' wattage capacities are reduced when put together in a single box. Dimmers produce heat and need air space around them in order to work properly. If too many dimmers are put too close together, they will overheat and cause problems down the line. Manufacturers' guidelines will tell you how much load a dimmer can handle when two or more dimmers are ganged together. A scored section along each side of the mounting plate or fin is designed to be snapped off for ease of mounting multiple controls in one enclosure.

4. Choose well-made dimmers. The rotary and toggle-style dimmers commonly available in hardware stores are poorly made. Don't expect much from a $4.95 dimmer; chances are that in six months they will fail.

5. Choose a dimmer style that matches the switch plates being used throughout the house. Some dimmers require special plates that aren't compatible with standard

Drawing 6.7
Dimmer switches come in many varieties. Here are a few for you and your clients to consider

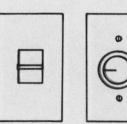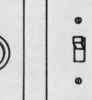

Slider Dimmer   Rotary Dimmer   Toggle Dimmer   Slider Dimmer   Touchtonic Dimmer   Preset Dimmer

plates for toggle switches or receptacles. Different plate styles in the same house may draw too much attention to an element that should be as low-key as possible.

Show your clients the choices available and let them select the style they like (See Drawing 6.7).

Preset Dimming Systems

Beyond standard dimmers are multiple-zone/multiple-scene controllers that allow for a great variety of predetermined light level combinations (See Drawing 6.8).

These systems, initially very expensive, have come down to a price range than can work within the budget of many projects.

Usually, these controllers are used only for entertaining spaces such as living rooms, dining rooms, entries and kitchens. Sometimes the master bedroom is included, as well.

For loads up to 2000 watts total in a room, there are a good number of 2-6 channel controllers that fit into a standard 4-gang box and use standard wiring.

For loads beyond 2000 watts, a remote dimmer panel is required. Then the loads are virtually limitless. The cost goes up significantly as more sophisticated components are added.

Most of these preset controllers have adjustable fade rates that help soften the transition from one scene to the next.

There are also computer terminal managed controllers, allowing people to turn on lights and even the hot tub from their car phones. For some, this is the ultimate dream package, but it certainly could scare the au pair if she's home alone.

Again, don't over-design the control system. A young couple, who are computer literate may love these bells and whistles. Older clients or young children may never get the hang of a sophisticated system.

Some commonly asked questions:

1. Do dimmers save energy? Yes, a fixture on a dimmer uses only the energy consumed.

2. Does using a dimmer shorten the life of the lamp? No, heat decreases the life of the lamp. Lower wattage consumption

Note: Some low voltage electronic dimmers have only a 300-watt capacity as compared to a standard 600-watt low-voltage or line-voltage dimmer. So care should be taken in the number of total watts each dimmer can handle.

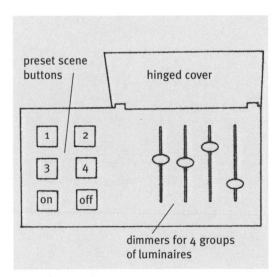

preset scene buttons

hinged cover

1   2

3   4

on   off

dimmers for 4 groups of luminaires

Drawing 6.8
A preset controller allows lights in a room to come on together at different levels called "scenes".

It is an industry recommendation that fluorescent luminaires with dimming ballasts should be operated at full capacity for 100 hours before dimming them, in order to break in the system. Some dimming ballasts are pre-cured by the manufacturer prior to being sold.

means less heat and longer lamp life. With Halogen lamps, it may be necessary to operate them at maximum illumination periodically to burn off any black residue which results from the tungsten evaporation.

3. Why does the dimmer feel warm to the touch? This is normal. It is the dissipated heat from the inefficient part of the control.

4. What about feedback from radios or stereos? With magnetic dimmers some buzzing may occur. Filters are usually installed in the better grade of dimmers. Remote "debuzzing" coils may be installed.

Central Home Lighting Control System

A logical step in central design was a system that allowed for remote control of the entire house from a few choice locations, such as the master bedroom and the door into the house from the garage. Options include turning all the lights in the house on or off, activating pathways of illumination, or turning on exterior security lighting (see Drawing 6.9).

A controller can override other switching and dimming devices in the house and change your clients' preset light levels. These systems involve complex wiring, so they may be cost-effective only in new construction. Remodel projects may require too much opening of walls and ceilings to justify the total expense.

Designers should review material and installation costs prior to starting the project. Gathering this information will help the clients make informed decisions.

A lighting designer, lighting showroom or electrical distributor can work with you and the contractor to determine the components required and help put the package together. You and your clients will greatly benefit from taking advantage of this service.

Drawing 6.9
A master control allows for activating any of the lights in the house from a central location

Low-Voltage Dimming

Low-voltage dimmer selection is crucial. Noise from the dimmer and transformer can be greatly reduced if these components are designed to work with each other. The manufacturer's specifications will help you make the correct choice.

A common mistake is specifying a low-voltage dimmer that is not compatible with the type of transformer used in the low-voltage luminaire. Find out the type of transformer before specifying the dimmer. If the information isn't readily apparent in the catalog, call your distributor to find out the information you need.

Flourescent Dimming

Dimming of fluorescents has come a long way. As people become more comfortable with fluorescent lighting, their demand for control of the light levels will increase as well.

First of all, in order to dim fluorescents, a special dimming ballast in the luminaire must be used. Standard ballasts will not work.

One of the recurring complaints about fluorescent luminaires was the incessant hum. Most of this came from magnetic ballasts that vibrate. This vibration may resonate against the metal luminaire housing, increasing the level of the hum. However, electronic ballasts are extremely quiet.

When you add a dimmer for a fluorescent luminaire, the electrician will need to run one or two additional wires from the dimmer to the ballast, and from there to the luminaire. Fluorescent dimmers require that there be additional wires to carry the dimmed current. For new construction, this would not be a problem. It is a bit more difficult for a remodeling project, not impossible, just more difficult. It's no more involved than adding the additional wiring for a three-way switch.

Here are some comparisons between dimming fluorescent luminaires with magnetic ballasts or electronic ballasts.

1. Hum is more evident in magnetic ballasts.

2. Both electronic and magnetic ballasts in the moderate price range are dimmable to 10-20% of the light output. Magnetic fluorescent dimmers usually have stops that keep users from dimming beyond the 80% mark.

3. High-end electronic ballasts can have full-range dimming capabilities and, in some instances, such as the Hi-Lume electronic dimming ballast by Lutron, allow for dimming unlike lamp-lengths together.

4. There are now magnetic ballasts on the market that offer full-range dimming as well. Both Bash Theatrical Lighting and Lighting & Electronics offer a wide selection of ballasts.

5. Fluorescent dimming can be pricey. But, considering the savings in electricity, the payback period can be very rapid - usually only two or three years in a typical household. The quality of the components is also a factor. Inexpensive fluorescent sockets can make dimming at low levels inconsistent and cause the light to flutter.

Some more reasonably-priced fluorescent dimming systems allow for dimming down to 50%. For some clients with budget constraints, this may be adequate.

6. Both magnetically and electronically ballasted fluorescent luminaires can be dimmed automatically, when tied to a photo-sensor, motion sensor, or time clock. This helps balance the need for artificial light with available daylight, or with use frequency. A home office would be a good candidate for this type of control system.

H.I.D. Dimming

The dimming technology for High Intensity Discharge lamps is still in the early development stages. At best, these sources can be dimmed only to 40% without shortening lamp or ballast life.

There is another system that dims the lamps down to 12-15%, but which shortens the life of both lamp and ballast.

The color shift of dimmed H.I.D. sources is generally undesirable.

As with much lighting technology, improvements in H.I.D. dimming should arrive in the next three to five years. Like all aspects of your lighting design, involve the clients: give them their options, and help them make an informed decision on how they want to (and can afford to) control the system.

Chapter Seven

SPECIAL EFFECTS

Don't let visually powerful lighting cause other aspects of the design to suffer loss of impact, unless the lighting is to be the central focus.

Now that you have worked your way through the basics of lighting design, this chapter will describe some relatively new products and lighting design ideas.

Homeowners might consider using special lighting treatments that are subtle or controllable enough to be experienced every day. Day to day use of specialized lighting, such as neon, fiberoptics, and framing projectors in residential applications may be too intense for your clients. Commercial spaces can be more adventurous, because they are visited less frequently by the same people. This approach is similar to choosing art: a particular painting or sculpture may have an impressive impact initially, but will your client love it enough to have it as a constant part of their daily existence? Just as in film music and computer graphics, people can get too creative, causing their creation to outshine the main event. Lighting special effects, can divert attention away from the rest of the space and the people in it. Still, working with your client to make a strong statement is certainly a design option.

Usually, lighting works best as part of the background, bringing people's attention to the environment, but rarely to the lighting itself. Designers put together many elements to create an overall effect. Don't let strong lighting cause other aspects of the design to suffer loss of focus or visual inclusion, unless the lighting is to be the central focus.

As in real estate, there are three important things to consider: location, location, location. Since these special effects can have such a strong presence, picking the right placement is a major factor. A glowing visual treat at the end of a long hallway pulls people towards it. Special effects can help give subtle direction to guests.

The interior entrance gives a first impression of the owners and of what the rest of the house is like. A special effect here can give a great visual clue as to what's in store.

Another property of special effects lighting is that it can give the designer the ability to add dimension to a space. Since the effect is often the strongest visual presence in the room, balancing its light level with the other lighting in a room will help add an intriguing 3-D feel.

Additionally, a visually-strong special effect can draw people's attention away from the less desirable elements of a structure, or possibly turn those undesirable elements into a fascinating design look. A room with a jumble of architectural features might benefit from the eye-catching characteristics of a special lighting effect.

Apply this concept to an outside situation: An illuminated sculpture could visually allow your client's neighbor's scruffy backyard to fall into the shadows. This involves using the glare factor as a positive aspect of a special effect. As your eyes adjust to the brightness of a neon sculpture, for example, your irises contract to accommodate that brightness. The surrounding area appears to darken considerably, at least as we perceive it. This helps the less-than-pleasant aspects of the neighbor's property to fade away into the darkness.

Framing Projectors

Also known as *optical projectors*, these luminaires can focus light to match the shape of the art or table which it is illuminating. The simpler models on the market use a series of shutters to make the beam of light approximate the size of the art. The more sophisticated units use a custom-cut metal template to match the shape of the art (see Drawing 7.1). Ready-made templates are also available for both high and low-end optical projectors

Before MR16's, MR11's and PAR36's came along, framing projectors were the only way to get a controlled beam spread. Theatrical luminaires gained that ability decades ago, but were simply too bulky for residential use.

Like everything that's discussed in this book, framing projectors have their pros and cons.

Drawing 7.1 (left)
A framing projector uses a template or series of shutters to "cut" the light to the shape of the art

Drawing 7.2 (right)
A table that is lighted with a framing projector must stay in its exact position or a disturbing rim of light appears on the floor

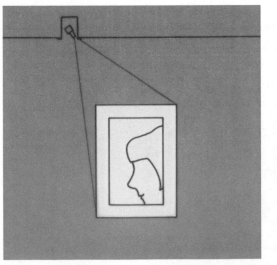
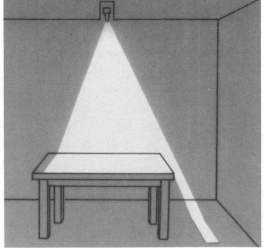

The expensive models do an excellent job of framing the art or sculpture. Inexpensive models have less precise optics, so the horizontal and vertical lines tend to be slightly convex.

The projectors can cause the art to predominate in the room. A recessed adjustable luminaire without an optical lens, using an MR16 lamp, would allow for some light spillage beyond the art piece itself. This actually helps integrate it into the overall design. Framing projectors can be adjusted to soften their focus to create the same effect.

Framing projectors may need regular adjustments if there is much movement in the building itself, such as kids running around, or if the house is built near a major thoroughfare. An out-of-adjustment framing projector pointed down onto a table can create a disconcerting corona of light on the floor (see Drawing 7.2).

The housing for the framing projector needs to be accessible. That means having a 12"-diameter cover plate in the ceiling or installing a trap door in the floor of the room above. If the room upstairs is an accessible attic space, there is no problem, but if it's a bedroom, then carpeting cannot be permanently installed there.

If your clients have a very valuable, prestigious piece of art with which they want to dazzle their family and friends, a framing projector would certainly do the trick. Sometimes, it's money that helps your client make the final decision. A good recessed framing projector by Wendelighting can cost around $850, plus the cost of installation. A simpler recessed adjustable MR16 luminaire would cost around $100, plus installation. Surface-mounted framing projectors can cost from $150 on up, and are less expensive to install than a recessed luminaire.

Recessed units are not very flexible. The lower-price units do an acceptable job of making a square, rectangle or circle, but usually cannot make irregular shapes. Higher priced framing projectors can mimic almost any shape, but must be refitted with a new template if the art is changed.

Neon

Neon has been around for a long time. It was very popular from the 1930's to the 1950's and then began a decline in the 1960's and 1970's. It is enjoying a resurgence of popularity, due mostly to craftpersons refining the art of neon and the more widespread acceptance by architects and designers. Also, the availability of improved transformers allows for a reduction in the noise levels, and easier dimming has made neon more user-friendly. Officially, neon in hard-

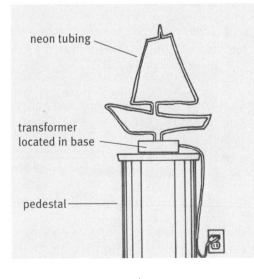

neon tubing

transformer located in base

pedestal

Drawing 7.3
A neon sculpture with a plug-in type transformer creates a strong visual statement

Courtesy of Fiberstars.

wire installations has not been allowed in residences for 40 years. The National Electrical Code does not allow systems of 1000 volts or more in residences. Neon normally runs in the 3000-15,000 volt range. But portable neon sculpture is legal and is now part of many private collections (see Drawing 7.3). The main precaution is to make sure that connections are secured - for protection from shock and to protect the tubing itself.

Neon is a glass vacuum tube filled with neon gas. When electricity is passed through the tube, the gas is excited to produce an orange-red glow. The addition of other gasses or phosphors, as well as the use of different-colored tubing, produces a variety of colors.

The glass tubing is heated and formed into letters or designs. The overall length and diameter of the tubing determines the size transformer that is needed. Systems with multiple tubes wired in series require more voltage.

Full brightness neon may be too strong to work comfortably within a residential design. Advanced dimming capabilities now offer a wide range of light levels so that the neon may be more easily integrated into a particular setting.

Helium and neon can be combined in a tube to create a curious luminous bubbling effect by using radio-frequency transformers. The bubbling effect is created as radio waves vary.

Cold Cathode

Cold cathode is a close relative of neon. The tubing is slightly larger (25mm) in diameter than neon (20mm or smaller). Basically, cold cathode is used for illumination purposes, whereas neon is used for signage and as an art form, although there is some crossover. Cold cathode, like neon, comes in more than 30 colors.

There are transformers that operate at less than the 1000 volt maximum so that they are able to be permanently installed in residences. Inspectors will want a UL (Underwriters' Laboratory), ETL, or other nationally-recognized testing laboratory listing.

In homes, cold cathode can do a wonderful job of indirect light behind cove or crown molding. Its small diameter, compared to standard fluorescent lamps, allows a less bulky architectural detail to conceal it (see Drawing 7.4).

The inside of a niche detail could also use cold cathode as its light source (see Drawing 7.5). It could also be used to outline the inside perimeter of a skylight (see Drawing 7.6).

Think of cold cathode as a source of light that isn't seen, but experienced, while neon is meant to be seen.

Fiber Optics

Fiber optics have come a long way in their application possibilities. Illuminated from one end, the light travels through the fiber optic filament or bundle of filaments in a tube.

One variety, called *end-lit* fiber optics, projects an intense light from the end opposite of the light source of the fiber-optic run. The other variety, *edge-lit* illuminates the length of the fiber-optic run. Presently, most fiber optic runs are illuminated by halogen or metal halide sources.

A fiber optic package usually contains two components: the fiber optic material itself and its source of light, usually called an *illuminator*. The illuminator is a remotely-located shielded box which houses a lamp, a ballast (if an H.I.D. source is being used), a fan (to keep the lamp cool and to help lengthen its life, and an optional color wheel or filter holder that allows the owner to change the color of the light output or allows for a gradual shift from one color to the next. Remember the 1950's when everyone had a silver metal Christmas tree and a motorized light that passed colored lenses in front of a spot light on a weighted base? Well, this is just a more refined version of the same principle.

Designers are often misled into assuming that edge-lit fiber optics are as bright as neon and are a viable substitute for neon or cold cathode on a project. While fiber optics and neon can be visually matched in brightness in certain environments, the true advantages of fiber optics are what they can do better than neon. Flexibility, ease of installation, and low maintenance have made fiber optics better suited for many functions usually reserved for other light sources.

Edge-lit fiber optics can do a good job of approximating the look of a neon sign or sculpture. One of its main advan-

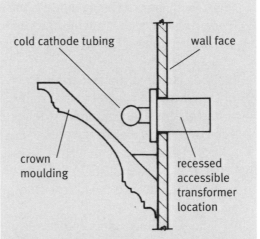

Drawing 7.4
Cold cathode could be mounted behind crown moulding to provide uplight (local codes need to be checked prior to installation).

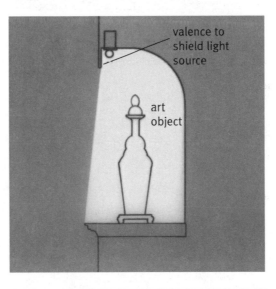

Drawing 7.5
Cold cathode can be used to light the inside of a niche (local codes need to be checked prior to installation).

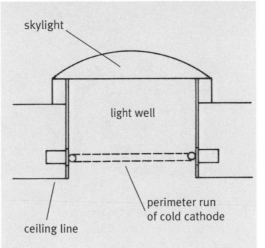

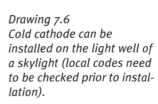

Drawing 7.6
Cold cathode can be installed on the light well of a skylight (local codes need to be checked prior to installation).

To keep the light level constant throughout the length of the run, you should keep the length of a fiberoptic run to 100 feet or less and loop the run back to the illuminator. Optionally, place an illuminator at both ends.

tages over neon or cold cathode is the ability to change the color at will. Red neon will always be red neon. Only replacement of the tube will give you a different color. While a simple change of lens is all that is required in fiber optics.

Another advantage is, since there is no electricity actually going through the tube, a fiber optic run can be integrated into water environments. How about your clients' signature or the outline of a giant fish in the bottom of their pool? You can inset a fiber optic run into the edge of the clients' pool or hot tub (see Drawing 7.7).

They can also be snaked through walls of glass block to add another dimension to that architectural detail at night (see Drawing 7.8).

You can illuminate a sandblasted acrylic hand rail from within by inserting an edge-lighted fiber optic bundle through the center, if it's a hollow straight run, or by fitting it into a rout on the underside, if it's a curved solid rail (see Drawing 7.9).

Fiber optics are dimmable, even the H.I.D. source are dimmable down to 40%. But since the output of illumination can be subtle, dimming is often not necessary.

You could edge-light a table, or a piano, or the steps leading to your clients' front door, or outline an architectural detail, or create an art piece.

Note: To keep the light level constant throughout length of the run, you should keep the length of a fiberoptic run to 100 feet or less and loop the run back to the illuminator. Optionally, place an illuminator at both ends.

End-lit fiber optics can perform a variety of interesting functions. For example, single strands of fiber optics can be pulled through a ceiling to create a starry night effect. The color wheel can do its trick of gradually changing the hue and intensity of those man-made constellations.

Drawing 7.9
A sandblasted acrylic handrail glows from within through the use of a fiber optic bundle running in a channel along the underside.

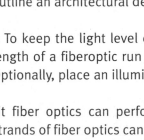

cross section of a sand-blasted acrylic hand rail

fiber-optic bundle

End-lit fiber optics (offered by Fiberoptic Lighting Inc., Niten Day, Industries & Drama Lighting, and others) are also being integrated into display lighting and downlights where remoting the light source is advantageous for maintenance purposes.

Fiber optics can be a very good way of adding a little sparkle or flexible accent light to an upcoming project. These examples are just a taste of what is possible.

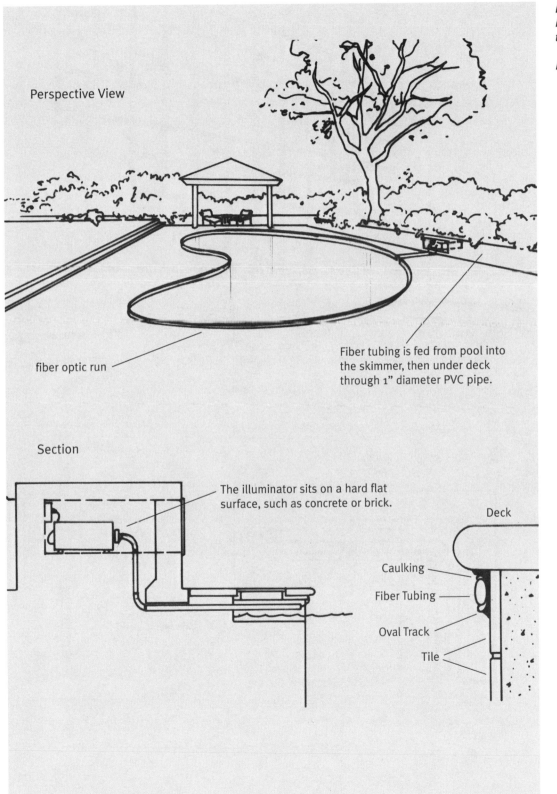

Drawing 7.7
Fiber optics can be used
to edge light a pool.

Design concept by Fiberstars

Perspective View

fiber optic run

Fiber tubing is fed from pool into
the skimmer, then under deck
through 1" diameter PVC pipe.

Section

The illuminator sits on a hard flat
surface, such as concrete or brick.

Deck

Caulking

Fiber Tubing

Oval Track

Tile

Drawing 7.8
Edge lighting of glass block
with fiber optics blends art
with architecture

Design concept by Fiberstars

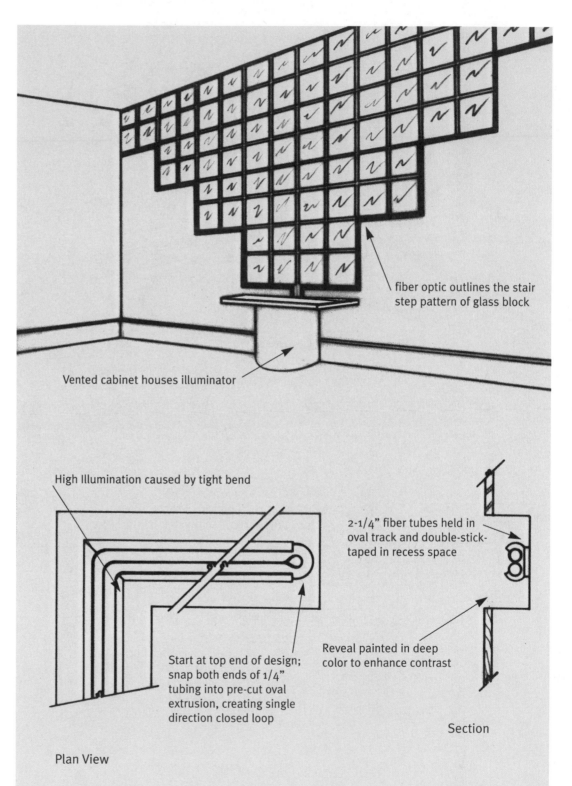

fiber optic outlines the stair
step pattern of glass block

Vented cabinet houses illuminator

High Illumination caused by tight bend

2-1/4" fiber tubes held in
oval track and double-stick-
taped in recess space

Reveal painted in deep
color to enhance contrast

Section

Start at top end of design;
snap both ends of 1/4"
tubing into pre-cut oval
extrusion, creating single
direction closed loop

Plan View

Section Two
Using Light

"' Incandescent' is how he thinks, but 'fluorescent' is what I want."

Chapter Eight

KITCHENS—THE NEW GATHERING PLACE

Kitchens have become the new centers for entertaining. The impact on lighting is that the kitchen should now be as inviting as the rest of the house.

Kitchens have become the new centers for entertaining. One reason for this is the change in the way we entertain. It's more casual than it was previously. As traditional roles were shared among family members, our approach to entertaining became more relaxed and interactive. Now it's more likely that guests will gather in the kitchen as the meal is being prepared, often helping.

New homes are being laid out by architects and builders to accommodate this change. As a result the trend is toward an *open plan* house, where the rooms blend together. The walls between the kitchen, dining room and family room have disappeared, this newly defined space is often referred to as a *great room*.

The impact on lighting is that the kitchen should now be as inviting as the rest of the house. It, too, must have controllable lighting levels, so that guests look good and feel as comfortable as in the other parts of the house. The color temperatures of the lamps should match, or at least be similar to, the color temperature in other areas of the house.

This definitely changes some of the lighting methods that have been around for a long time.

Sadly, we still see new kitchens, expensive kitchens, with a single source of illumination in the center of the room. Whether this is incandescent or fluorescent, it is essentially a *glare bomb* that provides little in the way of adequate task, ambient or accent lighting. As our eyes adjust to the glare, the rest of the kitchen seems even darker than it is. We see only the light source, and little of the surrounding room (see Drawing 8.1).

As you have learned, there is no single luminaire that can perform all the required functions. Here, as in almost everywhere else in the house, it is the layering of various light sources that create a comfortable, usable lighting design.

In the 1970's, builders seemed compelled to put a run of track in the center of the kitchen.

*Drawing 8.1
The illumination from a surface-mounted luminaire in
the center of the room is
blocked by one's own body.*

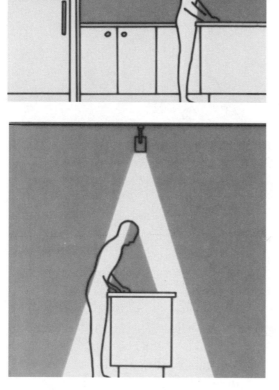

*Drawing 8.2
Track lighting is a poor
source of task light. Clients
will be forced into working
in their own shadows.*

*Drawing 8.3
Linear task lights
mounted towards the back
of the cabinet create a distracting glare when people
sit at the table.*

Track lighting is best used as a source of accent lighting. If you try to use it to light the inside of cabinets or down onto the countertops, your own head can get in the way, casting shadows onto the work surface (see Drawing 8.2).

The old surface-mounted luminaire in the middle of the ceiling can cause the same problems.

In the 80's there was a shift toward using a series of recessed downlights installed in a grid pattern in the ceiling. This was a little better, but still, by themselves, they cast harsh unflattering shadows on people's faces and your own head still eclipsed the work surface.

Undercabinet Lighting

The first step towards successfully layering the lights in the kitchen is the introduction of lighting mounted below the wall cabinets. This type of lighting provides an even level of illumination along the countertops. Since it comes between the work surface and your head, the lighting is much more shadow-free.

These linear task luminaires come in a great variety of styles and lamp sources. What we commonly see is a fluorescent strip luminaire mounted at the back of the cabinet. The drawback to this placement is that when people are sitting down in the breakfast area or the adjacent dining room, the lamps can hit them right in the eye (see Drawing 8.3).

What has been on the market for a while now are linear task lights, in incandescent or fluorescent versions, that mount towards the front of the cabinet. They project a portion of the illumination towards the back splash which then bounces light onto the work surfaces and out towards the center of the kitchen, without glare (see Drawing 8.4). This works well when the countertop is a material with a non-specular surface, such as a matte-finish plastic laminate, unpolished marble or Corian™.

Often lighting designers are faced with the challenge of lighting highly reflective countertops, such as polished black marble or glossy tile. These shiny surfaces act like mirrors, revealing the light source under the cabinets. In this situation, a solid reflector could be installed along the underside of the luminaire so that the light is directed only towards the backsplash.

Sometimes, in the worst case scenario, both countertop and backsplash have a gloss finish. This is a true lighting nightmare. If your clients are not willing to choose a surface with a matte finish, then the only solution would be to install miniature recessed adjustable luminaires with louvered faces. Inside the cabinets, a false bottom would need to be created in order to hide the luminaire housings (see Drawing 8.5).

Note: The incandescent linear task lights (including halogen and xenon) produce heat and can affect the items being stored inside of the cabinet. Items such as baker's chocolate can melt. It's a good idea to store perishables on an upper shelf if heat is a factor.

General Illumination

Ambient lighting plays an important role in the overall lighting design for the kitchen. It is this soft fill light that helps humanize the space.

There are a variety of ways to provide ambient lighting. One way, when dealing with nine-foot or higher ceilings, is to install a pendant-hung luminaire or a series of pendants along the centerline of the space. They can be made of an opaque material, such as plaster, or have a more translucent quality such as alabaster. Not only will they produce

Note: The incandescent linear task lights (including halogen and xenon) produce heat and can affect the items being stored inside of the cabinet. Items such as baker's chocolate can melt. It's a good idea to store perishables on an upper shelf if heat is a factor.

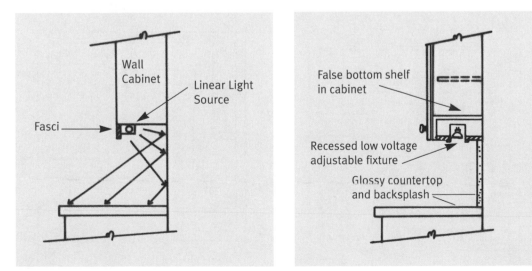

Wall Cabinet

Linear Light Source

Fasci

False bottom shelf in cabinet

Recessed low voltage adjustable fixture

Glossy countertop and backsplash

Drawing 8.4
A linear task light mounted at the front of the cabinet bounces illumination off the backsplash and onto the countertop.

Drawing 8.5
The only way to control glare when both the backsplash and the countertop are a gloss finish is to provide cross-illumination from recessed luminaire mounted inside the cabinets.

Drawing 8.6
A series of pendant luminaires can provide a wonderful, inviting ambient light.

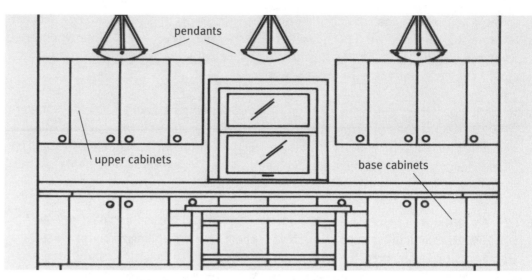

pendants

upper cabinets

base cabinets

Drawing 8.7
California's Title 24 states that either the center luminaire or the undercabinet lights must be fluorescent.

Drawing 8.8 (right)
California's Title 24 states that if there are two or more light sources in the room, the center luminaire must be fluorescent.

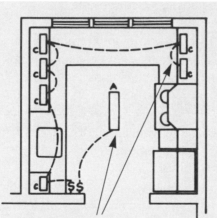

either the center light or the under cabinet lights can be fluorescent

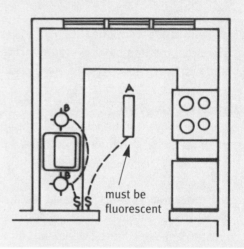

must be fluorescent

Drawing 8.9 (left)
California's Title 24 states that if there is only one luminaire in the room, it must be fluorescent.

Drawing 8.10 (right)
California's Title 24 states that if only undercabinet lights are used, then they must be fluorescent.

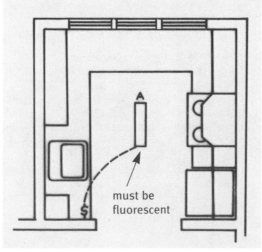

must be fluorescent

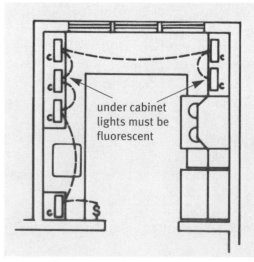

under cabinet lights must be fluorescent

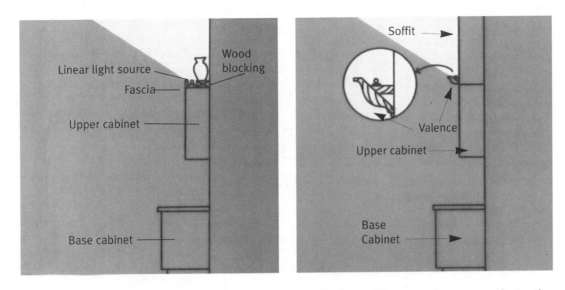

Linear light source

Fascia

Wood blocking

Upper cabinet

Base cabinet

Soffit

Valence

Upper cabinet

Base Cabinet

Drawing 8.11 (left)
Side section
Indirect lighting mounted on top of the cabinets not only produces ambient illumination, but could also highlight a client's collection.

Drawing 8.12
Side Section
If the wall is soffitted down to the tops of the cabinet, a valence of crown moulding could be installed to house a linear light source.

a wonderful ambient illumination, but they will also add a more human scale to the kitchen (see Drawing 8.6).

Another ambient lighting option is to mount linear luminaires above the cabinets, provided there is open space between them and the ceiling (see Drawing 8.11). This can be a wonderfully subtle way of producing the much-needed ambient light. If the cabinets don't have a deep enough reveal on top, a fascia can be added to hide the luminaires from view. Then put a shelf that is the same height as the fascia behind the light to create a potential display area along the top of the cabinet. Without the shelf, the display items would be visually cut off at the bottom.

The light source can be incandescent (including halogen and xenon) or fluorescent. Standard three-foot, four-foot and compact fluorescent tubes are easily and quietly dimmed using solid-state (electronic) dimming ballasts. This can also fulfill Title 24 requirements in California (see Drawings 8.7 – 8.10).

California requires the first switch in a kitchen or a bathroom to operate a fluorescent lamp and have an efficiency of at least 40 lumens per watt. Be sure to check local and state codes for building and energy regulations. California is not the only state to have codes geared towards energy efficiency. The emerging trend is to build energy conservation into building standards.

If the wall has already been bumped out or soffited above the cabinets, there is still an opportunity to build a cove or valence detail out of crown molding in which to house an indirect light source (see Drawing 8.12).

A less traditional approach would be to install a series of wall sconces on the face of the soffit instead of using a crown molding detail (see Drawing 8.13).

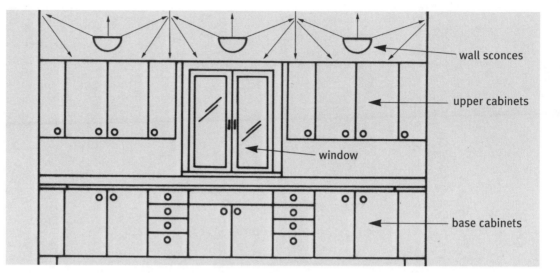

Drawing 8.13
Front view
A series of wall sconces
mounted on the soffit
above the overhead cabi-
nets will provide a modern
source of ambient
illumination.

Skylights can be great sources of general illumination, especially if they are made of a white opal acrylic material or are fitted with a diffusing material (refer to Chapter 5 on Daylighting). At night, light sources mounted within the light well can produce adequate fill light (see Drawing 8.14) while at the same time keeping the skylight from becoming a black hole after dark.

Don't let the fact that the kitchen you are designing is on the first floor of a three-story building stop you from installing a skylight. You can still install a *faux skylight* by opening up a recess in the ceiling (see Drawing 8.15), or using a ready-made faux skylight available from various manufacturers. This will create a feeling of increased height and openness.

Accent Lighting

The last consideration for the kitchen is accent lighting. Your client might have a few art pieces that can stand up to an occasional splash of marinara sauce. They deserve to be highlighted. This helps make the kitchen part of the overall open home plan. One tasty effect is to dim the ambient and task lights in the kitchen down to a glow, letting the accented art catch the attention once the party has moved to another area.

Many facets of your kitchen design will determine the way it is lighted. Not only do such variables as ceiling height, natural light and work surfaces affect the placement or amount of light used but there are other factors you should consider as well. Here is a checklist:

1. **Color**—Darker finished surfaces are more light absorptive. An all-white kitchen requires dramatically less light (40-50%) than a kitchen with dark wood cabinets and walls.

2. **Reflectance**—A highly polished countertop acts just like a mirror. Any under-cabinet lighting will show its reflection.

3. **Texture**—If your end design includes brick work or stucco, you might choose to show

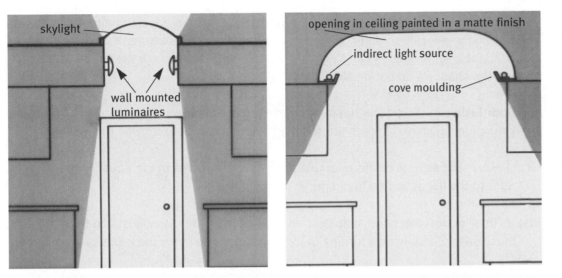

Drawing 8.14 (left)
Side Section
Light luminaires mounted on the inside of the light well can help provide ambient illumination for the kitchen at night.

Drawing 8.15 (right)
A faux skylight can be created to provide "daylight" even if the kitchen is on the first floor of a three story building.

off the textural quality of those surfaces. This is accomplished by directing light at an acute angle to the textured surface. Luminaires located too far away from the wall will smooth it out (which might be a good idea for bad drywall jobs).

4. **Mood**—Floor plans are more open now. Guests will flow from the living room to the kitchen to the dining room. The kitchen should be just as inviting as the rest of the house. Make sure that there is enough ambient light in the kitchen. This softens the lines on people's faces and creates a warm, inviting glow.

5. **Tone**—The warm end of the color spectrum works well with incandescent light, but cooler colors are adversely affected by the amber quality of incandescent light. Whites turn yellow and reds can turn orange. Make a color temperature choice that works well with skin tones and room colors.

6. **Code**—In California, designers must deal with Title 24 (the State Energy Commission's requirements for new construction and remodel work that exceeds 50% of the existing space). General lighting must be fluorescent (or have an efficacy of at least 40 lumens per watt) in kitchens and baths, and must be the first switch as you enter the room. Today many decorative luminaires are made to take compact fluorescent lamps, now dimmable in the quad versions. California is not the only state with such regulations. In the near future, most states will be affected (refer to drawings 8.8 - 8.11.)

7. **Windows**—Windows that let wonderful light stream in during the day while showing off landscaping will become black reflective holes at night, unless some thought is given to exterior lighting. Outside lighting will visually expand the interior space out into the exterior (see Chapter 15 on Landscape Lighting).

8. **Sloped Ceilings**—Even if there is enough space above a sloped ceiling to install recessed luminaires, special care must be taken to select units that don't glare.

9. **Pot Racks**—A pot rack may look just perfect over that center island on the plan, but it's extremely difficult to light a work surface through cookware. Consider recessed adjustable luminaires to cross light the surface or down through the center of the rack, but the shadows cannot be eliminated.

10. **Door Swings**—Make sure that switches are on the unhinged side of a door. Otherwise, your client will have to reach around the back of the door to turn on the lights.

Of course, many aspects of this checklist apply to other rooms in the house, so feel free to refer to the list as you go from area to area.

Just as interior designers and architects are pulling in kitchen specialists on their projects, those specialists are now turning to lighting experts to make the projects glow.

Since the lighting consultant specializes only in lighting, he or she has a vast library of catalogs and information relating solely to light and lighting products. The consultant can turn out a quality design in a short amount of time, and save the kitchen designer's time and ultimately the client cost in research hours.

Chapter Nine
ENTRANCES—SETTING THE TONE

Too often,
a foyer in a home
is restricted in
space, but with
lighting and
related design
techniques the
space can be sub-
tly transformed
into a vastly more
welcoming place.

Just as you judge a person by your first impression upon meeting them, you judge a home by what you see and how you feel at the entry. This makes it doubly important to set just the right mood and tone. The correct lighting is crucial in how people will respond, as well as where their attention will be drawn.

Outside, as evening guests approach a house, the lighting should provide an eye catching, welcoming feel, as well as security. You will also need to give specific cues showing which way to approach and enter. Be sure to light the house number. Chapter 15 discusses illumination for the outside in more detail.

Inside the home, ambient illumination should surround the entry with welcoming light. Fill light is especially important in this area. People need a gentle glow of illumination to help them feel at ease in a new surrounding. This complimentary light also allows the homeowners to look their best when greeting people. Good ambient light in the entry will help transform what is often an awkward moment into a more comfortable and enjoyable encounter.

Entry lighting can also energize a person's impression of a home. Highlighting a dramatic painting, sculpture or architectural detail can help guests feel at once both welcomed and impressed.

People respond to what they sense, not always to what is real. A room can be made to look large, airy and open, or cozy, when the reality may actually be quite the opposite. Too often, an entrance in a home is restricted in space, but with lighting and related design techniques the space can be subtly transformed into a vastly more welcoming place.

Illusion
Not all lighting solutions directly involve the use of lighting luminaires. Mirrors, for instance, can be used to create the illusion of greater space or light. Mirroring one wall

can make a room seem to expand in size. Mirrors will keep wall areas located farthest from the windows from falling into darkness or seeming less important.

Entries come in all shapes and sizes. Lighting can help you redefine the envelope of the space. Do you want the entry to look larger or more intimate? Do you want a look that dazzles, or a look of homey comfort? How about a combination of both? It is possible. For example, in a cramped entry area you can often use illusion and lighting to "steal" part of another room and visually make it a part of the entry area. In addition to mirrors, you might consider such techniques as using glass block, or directing an accent light onto a sculpture, flower arrangement or painting in a separate room adjacent to the entry to make that area seem like part of the entry.

Stairways in an entry area can provide additional opportunities for expansiveness. Illuminated, they make the room seem larger and provide another focus for the guest's attention upon entering the home. Lighting a painting mounted on the wall along the stairway, or illuminating plants or a sculpture on a stair landing, can also help a small entry assume the appearance of a grander entrance hall.

Switching and dimming systems can take the same entry that was made to look huge and dramatic for a big party or event and instantly transform it into a cozy and intimate greeting area for small friendly gatherings. Lighting *can* and *should* be that flexible.

With an integrated dimming system, your clients can create whatever kind of setting they'd like. There's nothing wrong with making the house seem foreboding when uninvited guests stop by, or having the lights go to full brightness when it's time for guests to depart.

Lighting directed toward the ceiling helps open up a space, and makes it feel larger and at the same time friendly and inviting. Illumination pointed down onto the floor makes an entry seem smaller because the darkened ceiling feels lower. Adding accent light to a darkened space creates highly dramatic settings. Accent light layered with ambient light provides a friendly environment with a bit of visual punch.

Too often entries end up looking unintentionally dark and uninviting, even if elegantly decorated. Usually it's because ambient light has not been considered. Either it was omitted entirely or the walls and ceiling may be too dark for indirect light to be reflected and diffused effectively throughout the space. If the lighting does not adequately light the ceiling, it not only makes the room seem small but neglects what often can be wonderful decorative elements in a design. Beams, coffers, moldings, ceiling frescoes and other well-lighted design components can become marvelous welcoming details, giving people something to engage their interest as they enter and expanding the space visually upwards. For parties, entrances can serve as auxiliary gathering spots when they are properly lit for people's comfort.

Planning

It's essential that you take lighting needs into consideration at the beginning of the design stage, because additions later will cost much more than adding appropriate lighting touches at the front end of a project.

Daylight also should be integrated into the design of an entry. Use available windows or add windows or skylights to provide some or all of the ambient light during daylight hours. (Recheck Chapter 5 on Daylighting for some application ideas.)

A common approach to lighting an entry is to put a decorative luminaire, such as a chandelier, in the center of a ceiling as the only source of illumination (see Drawing 9.1). If this is your design choice, then the bottom of an entry or hall pendant should be a minimum of 6'8" off the floor.

As a result, this one luminaire draws all the attention. Art, moldings, ceiling details and flower arrangements fall into secondary importance. Your clients, as they greet guests, will end up in silhouette, which does not give them a flattering appearance.

This is when "light layering" comes into play. For example, think about installing a source of ambient light, so that an existing chandelier can be dimmed to a subtle sparkle (see Drawing 9.2).

One option for the ambient light would be to install a pair of wall sconces flanking an art piece to provide the necessary glow of illumination. Translucent versions may draw too much attention to themselves; opaque-bottom sconces (made of metal, bisque or plaster) will cast light upwards, softening the shadows on faces and filling the entry with a pleasing glow of illumination.

Rule of Thumb — Mount opaque-bottom wall sconces above eye level; normally 6 to 6-1/2

Rule of Thumb — Mount opaque-bottom wall sconces above eye level; normally 6 to 6-1/2 feet above the finished floor. This applies to 8 or 9-foot ceilings. When working with a higher ceiling, the luminaire can be mounted higher.

Drawing 9.1 (left)
Using only a chandelier in the center of the ceiling draws too much attention to itself and the rest of the room and the people are not illuminated properly.

Drawing 9.2 (right)
Wall sconces provide the ambient light for the room, and allow the chandelier to be dimmed, so that it gives the illusion of providing the illumination, without attracting too much attention to itself.

*Drawing 9.3
Torchieres can also be a
good, easy way of adding
ambient light. They provide
the illumination that people
will think comes from the
chandelier, which can be
dimmed to an appropriate
level.*

*Drawing 9.4
Cove lighting inside the
moulding detail provides
ambient light without glare.
Again, the chandelier gives
the illusion of providing illumination, but is dimmed so
it will not produce too much
distracting brightness.*

*Drawing 9.5
A pendant-hung indirect
luminaire can be substitut-
ed for the chandelier, to
provide the necessary
ambient light. Candlestick
wall sconces could be
added for sparkle.*

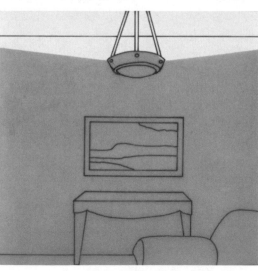

feet above the finished floor. This applies to 8 or 9-foot ceilings. When working with a higher ceiling, the luminaire can be mounted higher. Do not mount the luminaire closer than 2 feet from the ceiling; otherwise there'll be a hot spot above the sconce.

A second possible source of ambient lighting is the torchiere (see Drawing 9.3). If electrical outlets are already available, then an electrician is not needed. Torchieres are the quickest and easiest way of adding ambient light to a room. Remember to choose a luminaire with a solid reflector-type shade that provides uplight only. Otherwise attention will be drawn away from the other, more interesting, aspects of the entry. A half-switched receptacle would allow your clients to turn on the torchiere using a wall switch instead of turning it on at the luminaire itself (refer back to drawing 6.1 in Chapter 6.)

A more architecturally-integrated solution to the question of ambient light would be to install cove lighting (see Drawing 9.4). A linear light source can be hidden behind molding details to uplight the ceiling along the perimeter of the entry. In new construction, this is an inexpensive addition to overall building cost. In a remodel situation, this could be the costliest and most labor-intensive of the ambient lighting solutions.

Lastly, you could substitute an opaque-housing, pendant-hung, indirect luminaire in place of a chandelier to create the necessary fill light (see Drawing 9.5). This would eliminate the source of sparkle, so you could install translucent or candlestick-type wall sconces on either side of the art as the sparkle of light for the entry. Decorative wall sconces are normally mounted at 5-1/2' on center (center line of the junction box) above the finished floor.

Now that you have addressed the decorative and ambient light questions, you can tackle the accent lighting. While the common choice for accent lighting is apt to be track lighting, it is generally not the best solution. For some rooms it may be the only feasible choice, because of the architecture, construction complications, or cost. But track lighting visually intrudes into a room, and makes people feel as though they're on stage or on display.

When possible, recessed adjustable luminaires mounted in the ceiling are the better solution. These can be directed to particular spots that need highlighting, and are much more low profile. Even in an existing home, recessed luminaires, made especially for remodel, can often be installed within a reasonable budget and with a minimum of mess. Always hire a professional electrician to get the job done properly. A well-installed job makes you look good.

Can Halogen Lamps be Used as Downlights?

The big question here is: what is the best use of downlights? Try not to place downlights over seating areas. They cast hard shadows onto people's faces, making them look tired and older. Downlights should only be located over stationary objects such as plants, sculpture, tables, etc... The trend we are seeing in the industry is using recessed adjustable luminaires that allow people to redirect the lighting, as art and furniture is moved around. Interiors are no longer static the way they were in the 50's and 60's. Both MR16's and MR11's are halogen sources, and can be good sources of accent light.

So can halogen lamps be used for downlights? The answer is yes. However, whether you choose to use downlighting in a project becomes the more pertinent question. In magazines, we constantly see rooms filled with a series of recessed downlights casting light in circles on the floor. Their intended purpose is to provide ambient illumination. Downlights, no matter what lamp is used, are a poor source of fill light, because of the shadows they cast and the absence of light to reflect back toward the ceiling. Use other sources, such as wall sconces or cove lighting, for the ambient lighting. Use adjustable downlights for accent lighting. In extremely rare situations it might be necessary to use downlights where no other solution is possible (This does not mean a limited budget situation!).

Please, avoid specifying straight downlights. They don't give the flexibility needed to highlight different sizes, shapes and mounting heights. As was mentioned before, homeowners are not static with their interiors anymore. With today's transient lifestyle, flexibility is a necessity to accommodate new furnishings or new home owners. A straight downlight offers very few options and limits your clients' choices.

Try not to place downlights over seating areas. They cast hard shadows onto people's faces, making them look older.

Drawing 9.6 (left)
Weighted bases are avail-
able that can hold track
heads, so they can be used
as portable accent lights.

Drawing 9.7 (right)
Stake lights in planters can
cast intriguing shadows on
the walls and ceilings.

On the other hand, don't go crazy with recessed adjustable luminaires and fill the entire ceiling with holes, often referred to as the *planetarium*, or *swiss cheese* effect. Don't feel you have to light everything in a room.

Accent light does not have to be permanently installed. There are a number of portable luminaires, readily available, that will work in temporary, lower-budget, or historically-protected homes. In addition, most track lighting companies manufacture weighted bases to accommodate their trackheads, converting them into portable accent lights (see Drawing 9.6).

There are also "stake lights" that fit into planter pots to uplight plants and cast intriguing shadow patterns on the ceiling (see Drawing 9.7). This can add wonderful texture to a space. Refer back to Chapter 4 for other portable options.

Be sure to hide light sources as much as possible. Let what is being highlighted come into focus, not the luminaires themselves.

Use your assets, whatever they are; turn straw into gold. Well-designed lighting can be powerful alchemy in the entry and throughout the rest of the house.

Chapter Ten

LIVING ROOMS—LAYERING COMFORT WITH DRAMA

Lighting in the living room should be as flexible as the rest of the home's components, and it needs to be controllable enough to satisfy a variety of needs.

The way we view living rooms today is very different from the way many of us experienced them in our youth. Living rooms seemed to be "off limits". No one would enter unless company came. It was as if an invisible braided rope kept the family out. It's almost like the way people treat guest towels: "Too nice to use".

Not that it was a place we really wanted to go, anyway. The furniture was formal and uncomfortable. Sometimes it had plastic slipcovers that your legs stuck to in the summertime. Even the lampshades may have had plastic covers - the kind that looked like big shower caps. That "hands off" feeling is finally softening a bit. The furniture is getting more comfortable and their arrangements more relaxed.

Nowadays, families have reclaimed the living room. It's not just reserved for special occasions anymore. Furniture plans are less static. It used to be that everything stayed in exactly the same place. Paintings, sculpture and plants are now rotated around the house to keep the look fresh. The living room is often rearranged at holidays to accommodate a Christmas tree or Hanukkah display, or to set up a buffet for a Fourth of July feast.

Lighting should be as flexible as the rest of the home's components, and it needs to be controllable enough to satisfy a variety of needs.

Although entertaining at home has been very popular, it's still the owners of the house that spend the greatest amount of time there. Your first concern is to give them adequate illumination for their day-to-day activities. Then layer that with lighting options for entertaining.

Ask your clients what they plan on doing in the living room. Will they read there? Will they watch television? If they have children, will they want to do puzzles on the floor or board games on the coffee table? Getting a picture of how the space will be used will help you decide how to light it.

Ambient Light

The first concern is to create adequate ambient light. There are many ways to provide ambient light. The typical eight-foot ceiling offers the least number of options. One solution would be the installation of four wall sconces (see Drawing 10.1), mounting them 2 feet down from the ceiling.

A possible alternative would be a pair of torchieres flanking the fireplace, although the illumination would be less even than with the four wall sconces (see Drawing 10.2). If the room is large, consider using two torchieres at diagonals from each other.

Torchieres provide an excellent ambient light for a room. Their main job is to fill the volume of space with an overall illumination that softens the shadows on people's faces and shows off the architectural detailing. If you have a white or light colored ceiling, the torchiere can provide a suitable secondary task lighting for reading. It should be used only for light reading like newspapers and magazines. For serious reading (books, magazines) we recommend the use of pharmacy-type lamps that position the light between your head and the work surface.

If you have a living room with nine-foot or higher ceilings, you have more open options.

A pair of pendant luminaires with an overall length of two feet to thirty inches would work well for a nine to twelve-foot flat ceiling in a room 15 x 15 (see Drawing 10.3). A pitched ceiling would require luminaires adapted for the slope.

A higher ceiling would also work well in conjunction with a cove lighting detail, where the light source is hidden behind a crown molding or valence detail (see Drawing 10.4).

Living rooms with gabled ceilings and support beams that are parallel to the floor offer an additional option. In this situation, linear strip lighting can be mounted on top of the

Drawing 10.1 (left) Floor plan: A series of four wall sconces would be a viable solution to provide adequate ambient light for this living room.

Drawing 10.2 (right) Two torchieres can also provide the room's ambient light, but it will not be as evenly lighted as when four wall sconces are used.

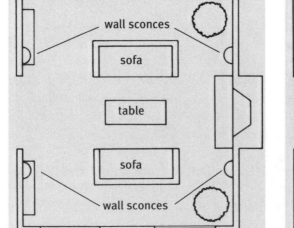

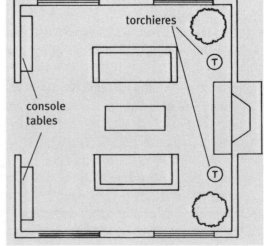

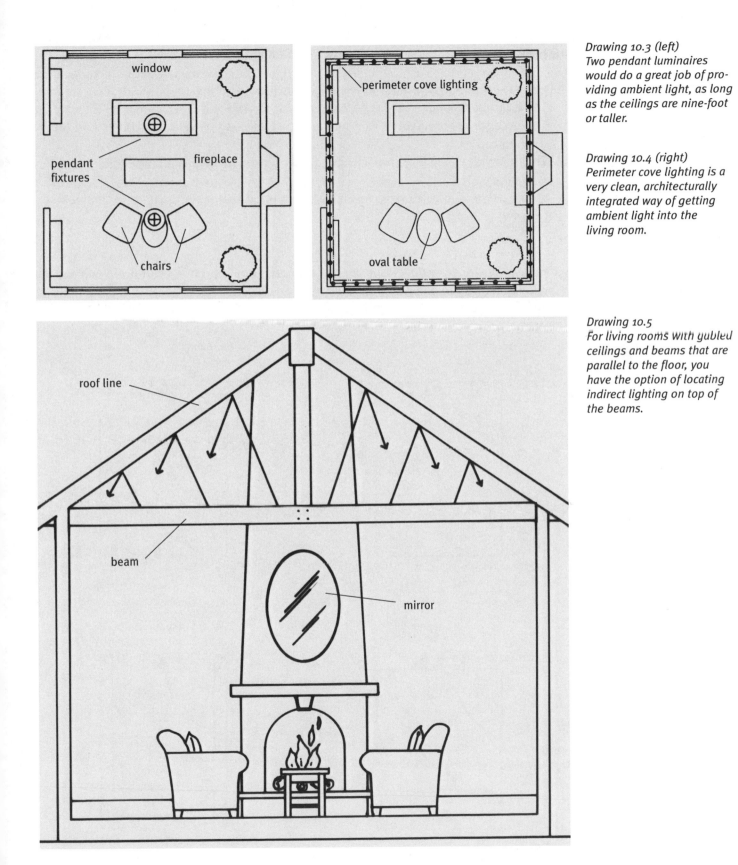

Drawing 10.3 (left)
Two pendant luminaires would do a great job of providing ambient light, as long as the ceilings are nine-foot or taller.

Drawing 10.4 (right)
Perimeter cove lighting is a very clean, architecturally integrated way of getting ambient light into the living room.

Drawing 10.5
For living rooms with gabled ceilings and beams that are parallel to the floor, you have the option of locating indirect lighting on top of the beams.

beams to provide the ambient light from a totally hidden source (see Drawing 10.5).

There are four different methods of installing this linear lighting.
1. The first way would be to rout a channel in the top of the beam.
2. The second way to hide the light luminaire would be to place it in a channel on top of the beam.
3. The third way would be to run a length of quarter-round molding along either side of the luminaire.
4. The fourth would be to wrap fascia boards around the beam to create a channel (see Drawing 10.6).

Symmetrically-sloped ceilings are becoming more common architectural elements. This results in one very tall wall that often becomes a dead space, because it is too high to

Drawing 10.6
Here are four methods of masking the linear lights mounted on top of vertical support beams.

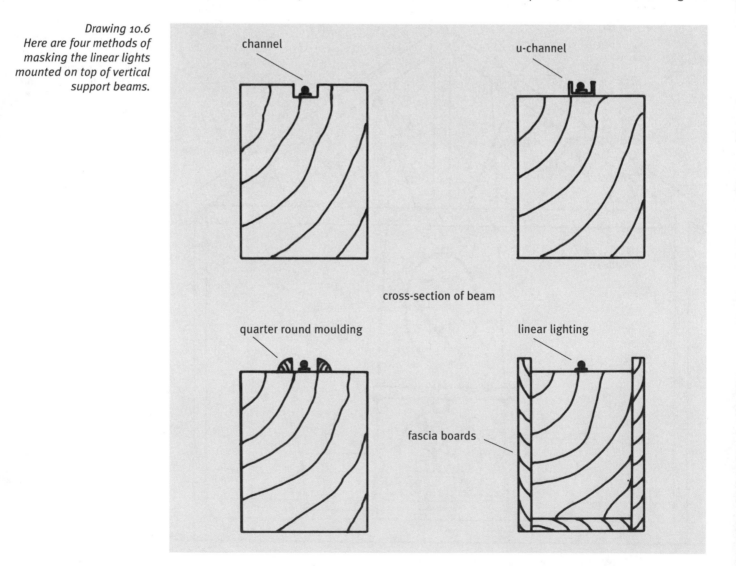

channel

u-channel

cross-section of beam

quarter round moulding

linear lighting

fascia boards

hang art. It is a perfect spot to mount a series of wall sconces that fill the room with an abundant amount of ambient illumination (see Drawing 10.7).

Accent Lights

Once you have decided what method you'll use to give the living room its fill light, the next step is to decide on accent lighting.

The type of luminaire you choose for the source of accent lighting needs to be flexible. As your clients move furniture and art around, the lighting needs to accommodate each new arrangement.

Remember, when straight downlights are used for accent lighting they offer no flexibility at all. If the highlighted object is moved, the clients are left with a circle of light on the floor.

If you specify recessed adjustable luminaires, you will provide clients with the adaptability they need. There are both line voltage and low voltage versions of these luminaires (refer back to Chapter 4 for comparisons between these two options). One of the advantages of a low voltage recessed adjustable luminaire is the ability to use a smaller aperture trim in the ceiling, which draws less attention to the unit itself.

If you are working on an existing home which already has recessed luminaires, it is possible to leave the housings (the main part of a recessed luminaire installed inside the ceiling) and replace the trims (the visible part of a recessed unit, attached to the housing) with line or low voltage adjustable versions. In new construction and remodel projects, placement of these accent lights is dependent on what is to be highlighted. That's why it's so important to know the furniture plan before laying out the lighting.

Drawing 10.8 shows a living room with the furniture centered around the fireplace. The

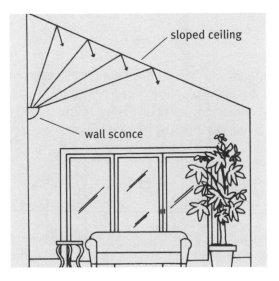

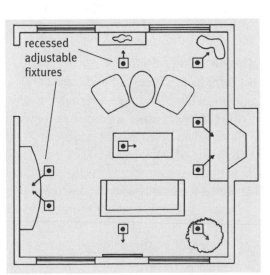

Drawing 10.7 (left)
Make use of the tall wall by installing a series of wall sconces for the much-needed ambient illumination.

Drawing 10.8 (right)
This is a partial lighting plan showing recessed adjustable luminaires used for accent lighting.

Angle of reflectance— The angle at which a light source hits a specular reflective surface equals the angle at which the resulting glare is reflected back.

recessed adjustable accent lights are positioned to highlight various components of the overall design.

On the north wall, the sculpture on the table, located between the two windows, is illuminated with a single recessed adjustable luminaire.

On the east and west walls, two pairs of recessed adjustable luminaires cross-illuminate the art over the console and the fireplace. It's better to use two luminaires to light a painting that has a glass or Plexiglas™ face, because a single accent light centered on the art could simply reflect back into people's eyes. By cross-illuminating, light is directed away from the normal viewing angle. The angle of reflectance plays a role whether one, two, or more luminaires are used.

Angle of reflectance—The angle at which a light source hits a specular reflective surface equals the angle at which the resulting glare is reflected back.

The single recessed adjustable accent light on the south and east walls illuminate objects on a small table, a plant, and a corner sculpture. The luminaire in the center highlights the coffee table. The flexibility of recessed adjustable accent lights allows the coffee table to be illuminated even though the luminaire isn't centered over the table itself.

Track Lighting

Sometimes track lighting ends up being the design choice when other options aren't available. For example, if there were inadequate ceiling depth in which to house recessed luminaires, a surface-mounted system must be used.

If track lighting is your choice, a good arrangement in the living room would be to run it around the perimeter (see Drawing 10.9).

You can make the track seem more architecturally integrated by installing a run of molding on either side of the track (see Drawing 10.10).

Halogen Bridge Systems

An alternative to recessed and track lighting is a relatively new product, generically called a *halogen bridge* or *two-wire* system. This is a low voltage set-up of two wires strung parallel to each other across a ceiling space. Luminaires, normally using MR16 or MR11 lamps, can be clipped and locked into place along the wires to highlight various objects below (see Drawing 10.11).

This works especially well in homes where the ceilings are sloped; otherwise, changing the lamps in recessed or track luminaires would be difficult because of their inaccessibility.

Note: Even though these luminaires are low in wattage they are high in amperage, requiring a specific gauge of wire. Make sure the installing electrician knows this and uses the

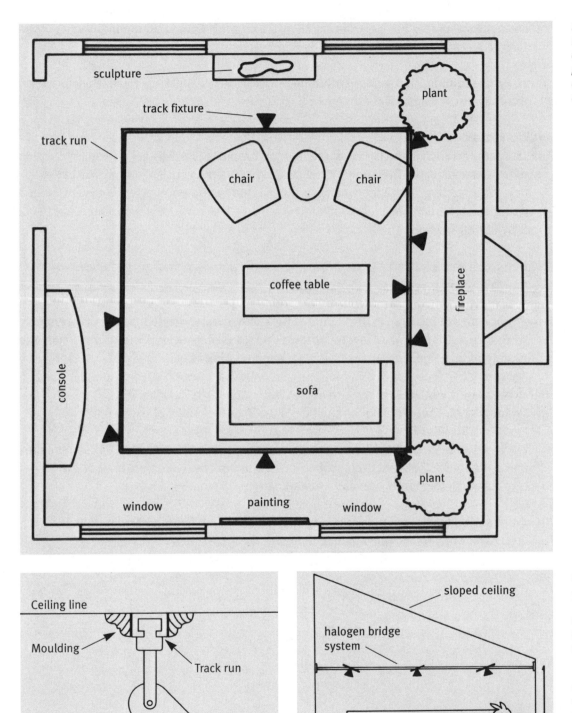

sculpture

track fixture

track run

chair

chair

plant

coffee table

fireplace

console

sofa

window

painting

window

plant

Drawing 10.9
If recessed luminaires aren't a possibility on a particular project, track lights set up in a perimeter pattern would do an acceptable job.

Ceiling line

Moulding

Track run

Track head
(lampholder)

sloped ceiling

halogen bridge
system

Drawing 10.10 (left)
Side section
The addition of mouldings on either side of the track run will help it blend in a little more architecturally.

Drawing 10.11 (right)
Using a "halogen bridge" system lets the luminaires be reached more easily, to change bulbs or readjust the lights.

correct wire gauge. The high amperage also limits the number of luminaires per transformer. Normally the maximum load is 300 watts, or six 50-watt lamps.

Normally, the lighting showroom distributing the wire system, will help you or the electrician put the correct components together.

Task Lighting

The last consideration in the layout of the lighting design is task lighting. The main function of task lighting in the living room is for reading or related activities. Since the best position for a reading light is between one's head and where one's attention is focused, a pharmacy-type luminaire or a tabletop luminaire will work very well (see Chapter 4, Drawing 4.3).

The bottom of the shade on a floor or table luminaire should be 40" to 42" above the floor to avoid glare.

If the floor plan has furniture in the middle of the room, it's a good idea to specify floor plugs so that cords don't cross the floor to a wall outlet (see Drawing 10.12). Crawl space or basement accessibility is a necessity in a remodel project.

In this case study, both sofas have a switched floor receptacle in which to plug the pharmacy luminaires. These luminaires can be placed on either sides of each sofa.

The west and south walls have half-switched receptacles, so that possible table luminaires and uplights for the plants can come on with the flip of a switch, instead of the homeowner having to go from luminaire to luminaire.

Remember, it's all the elements of lighting (task, ambient and accent) that must be considered to create a totally functional and adaptable lighting design.

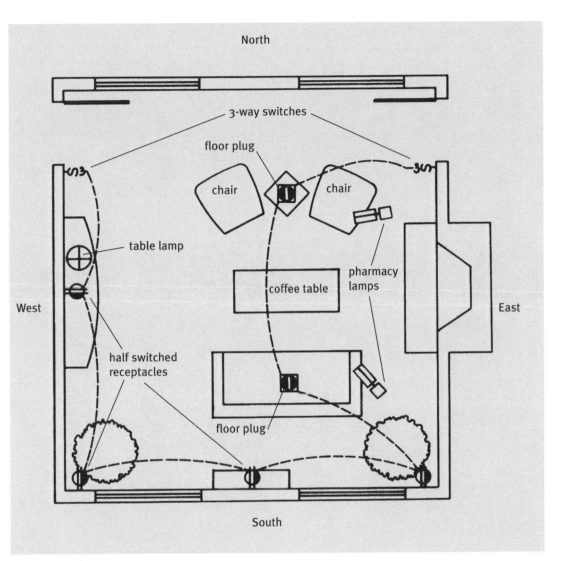

North

3-way switches

floor plug

chair

chair

table lamp

pharmacy
lamps

West

coffee table

East

half switched
receptacles

floor plug

South

*Drawing 10.12
Switched floor plugs help
keep cords from being
tripped over.*

Chapter Eleven

DINING ROOMS—THE MAIN EVENT

Dining rooms have been transformed in some homes into multi-use spaces. These changes set up a need for adjustable lighting and have forced homeowners and designers to rethink the use of a chandelier.

Dining rooms used to be the last holdout for a traditional static furniture arrangement. Now they, too, have been transformed in some homes into multi-use spaces. Homeowners realized how little of their time was spent actually entertaining, and they wanted to reclaim this often under-utilized room for additional purposes.

Previously, the dining room table itself was always too big for anything other than a meal for eight or ten people. Tables have become more flexible in size, folding down to more intimate seating for four, or divided in two to make a pair of game tables. Even the homeowners that kept their large table want to be able to push it against a wall for buffet dining.

All of these changes set up a need for adjustable lighting *and* forced homeowners and designers to rethink the use of a chandelier.

Chandeliers

For eons, the dining room table has been centered under the chandelier. Many people spent countless hours of their lives making sure that this alignment was just perfect.

As dining room tables began to move around, the chandelier in the center of the space started getting in the way. For those clients who want a traditional feel, but with the flexibility to create a more multi-functional room, there are a good number of choices:

1. Specify a decorative luminaire that hugs the ceiling so it doesn't look odd when the table isn't in the center of the room (see Drawing 11.1).

2. Select a pendant light on a pulley system that allows your client to raise or lower the luminaire. There are a good number of European-designed pendants available, primarily contemporary styles (see Drawing 11.2).

*Drawing 11.1 (left)
A decorative luminaire
mounted tightly to the ceiling will be less distracting
when the table is
moved aside.*

*Drawing 11.2 (right)
A pendant light on a pulley
system can be raised out of
the way when the table
is moved.*

*Drawing 11.3 (left)
A dome detail in the ceiling
will allow for a more traditional hanging chandelier
without hitting people in
the head when the table
is relocated.*

*Drawing 11.4 (right)
A table with no chandelier
makes use of three
recessed adjustable lumi-
naires to highlight the table
setting, as well as
the centerpiece.*

3. A traditional multi-armed chandelier of brass or crystal could be hung in a recessed dome so it's visual relationship is linked to the ceiling configuration rather than the table location (see Drawing 11.3). In a remodel project where it is too expensive or there is inadequate attic space for a dome, a decorative ceiling medallion will create a similar illusion.

The luminaire over the table should be at least 12" narrower than the tabletop and the bottom of the fixture should be 30" to 36" from the top of the table. If the ceiling is higher then 8', add 3" per each foot over the 8' height, but this can vary depending on the layout of the room.

4. Many clients are foregoing any decorative luminaire at all and are relying instead on recessed adjustable luminaires to provide illumination for the table, no matter where the table is placed. In Drawing 11.4, three recessed adjustable luminaires are used. The

middle one highlights a flower arrangement in the center of the table. The two outside luminaires cross-illuminate the tabletop itself, adding sparkle to the dishes and silverware. Make sure the two outside luminaires are not pointed straight down: this would cast harsh shadows on the people at both ends of the table and create glare from the reflective top of the table itself.

Keep these two luminaires pointed at an angle from the vertical that is less than 45%. At 45%, the light may glare into people's eyes. If luminaires are aimed at an angle less than 45% light will first hit the tabletop giving a soft complimentary underlighting of people's faces.

Drawing 11.5 shows how these same luminaires can be redirected towards the wall when the table is being used for a buffet.

For those clients happy with a traditional setting in their dining room, the addition of two recessed adjustable lights on either side of a chandelier will help add drama to the table. This allows the chandelier to be dimmed to a pleasing glow, while giving the impression of providing the table's illumination (see Drawing 11.6).

A large chandelier may require additional support at the junction box. Most standard junction boxes will support up to 50 lbs.

Large chandeliers installed in high ceiling areas may require a pulley mechanism mounted above the ceiling to lower the chandelier for ease of re-lamping and cleaning.

Ambient Illumination

Whichever solution you choose for the table illumination, this alone will not complete the lighting scenario. Ambient light, and additional accent lighting, should still be considered. While it's true that the decorative luminaires will provide some illumination for the

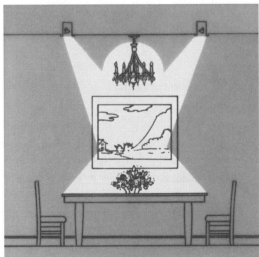

Drawing 11.5 (left)
Recessed adjustable luminaires over the dining room table can be redirected to light the table when it is against the wall serving as a buffet.

Drawing 11.6 (right)
Two recessed adjustable luminaires in combination with a chandelier give highlighting to the table and let the decorative luminaires take all the credit.

dining room, they can easily overpower the rest of the elements in the space if turned up too brightly. This is the same problem that occurs in entrances (see Drawing 9.1, Chapter 9).

Adding ambient lighting is relatively straight forward. Most of the options mentioned in previous chapters work here as well: torchieres, wall sconces and cove lighting. If the dining room you are working on has a dome detail, the perimeter could be illuminated so that the fill light is bounced off the dome's interior. If it is a beamed ceiling, channels can be used, as discussed in Chapter 10 on living rooms (See Drawings 10.5 and 10.6, Chapter 10).

Accent

The recessed adjustable luminaires over the dining room table already address accent light for the table itself, and the centerpiece. Downlights in chandeliers may be used to provide accent light for centerpieces. The next spaces to look at are the walls, side table or buffets, and plants.

For art on the wall, don't feel that every piece has to be illuminated. It's all right to let some pieces fall into secondary importance. It lets them be "discovered" as guests take a second look around the room.

Add one or two recessed adjustable luminaires to accent the side table, buffet, or console. A silver tea service will sparkle and a buffet dinner will look even more scrumptious when highlighted.

Plants can be uplighted, downlighted, or both. Broad-leaf plants like fiddle-leaf figs are better illuminated from above or backlighted. More airy-leafed plants, such as ficus, can be illuminated from the front, casting leaf patterns on the walls and floor. They can also be uplighted, which creates a pattern on the walls and ceiling. Palms are best shown off when they are lighted from both the top and from below. The sculptural quality of cactus calls for lighting from the front at a 45% angle and off to one side to help add dimension.

Candles

Candles should be used correctly as well. No light source goes undiscussed in this book!

Typically, at the dinner table, you artfully place candlesticks around the centerpiece. When you and your guests sit down at the table, that candle flame is right at eye level. When you look at the flame for a while, then towards the guests, you'll notice that there is a black hole where their heads used to be - much like the effect you experience after someone has taken a flash picture of you. To solve this problem, use candles that are either lower or higher than eye level. That way, you'll get that soft golden glow but the candle will not distract from looking at the person across. Reviewing Drawings 11.7, 11.8, 11.9 should give you an idea of what can happen.

Take a look at Drawings 11.10 and 11.11 to look at one way a traditional dining room layout could be lighted.

This particular design used a traditional chandelier flanked by two recessed adjustable luminaires (as shown in Drawing 11.6).

Four wall sconces are being used to provide the fill light, while five additional recessed adjustable luminaires have been installed to highlight the art and tabletops.

Drawings 11.12 and 11.13 show how a less traditional dining room might be illuminated. The table is actually two 4-foot by 4-foot tables that can be separated for smaller family dinners, card games, or even just sorting through bills.

The use of three recessed adjustable luminaires lets the lighting follow the tables wherever they end up in the room. This time, a pair of torchieres on the east wall are the sources of ambient light. Seven additional recessed adjustable luminaires accent the art above the fireplace, the sculpture on the west wall, and the two plants in the northwest and southwest corners of the room.

As in the rest of the house, *light layering* is the key. The addition of dimmers will help bring out the true beauty of the chandelier, while allowing for the full brightness needed to clear the table or work a puzzle. The goal is to create a balanced illumination that enhances the other elements of your design.

Drawing 11.7
Standard-height candles obstruct your view of the person across the table.

Drawing 11.8
A low votive candle provides a soft, complimentary uplighting

Drawing 11.9
A tall candle adds the romance without the glare

Drawing 11.10
This is a possible lighting layout for a more traditional dining room furniture plan

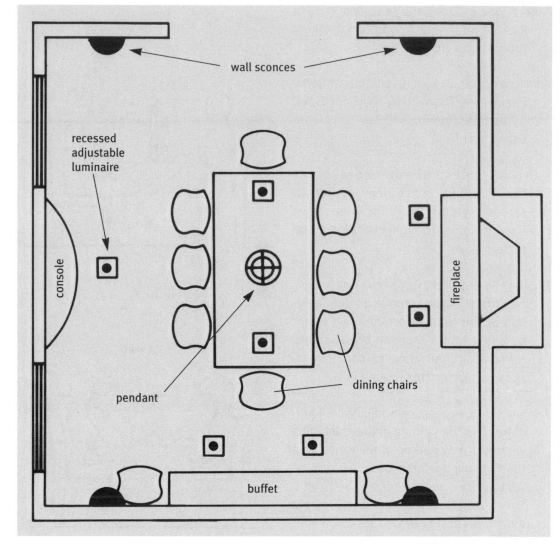

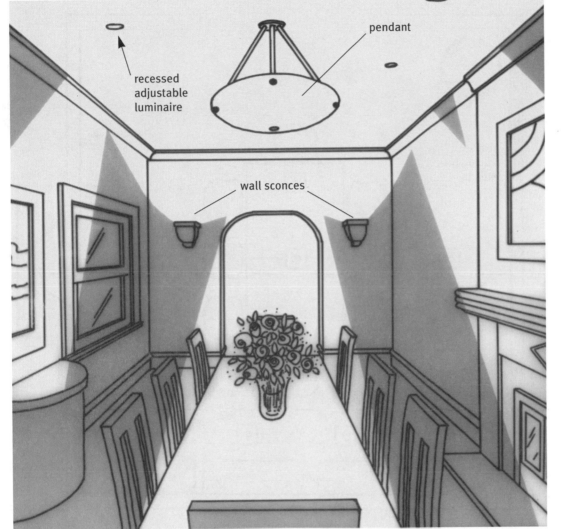

recessed
adjustable
luminaire

pendant

wall sconces

Drawing 11.11
This perspective drawing
shows what the room looks
like when translated from
the lighting plan (shown
on Drawing 11.10).

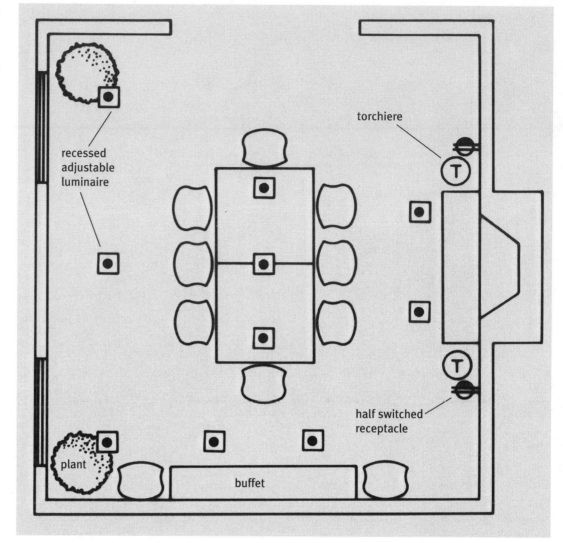

*Drawing 11.12
This is a more non-traditional furniture plan.
Note that the table can be separated to make smaller tables.*

torchiere

recessed adjustable luminaire

T

T

half switched receptacle

plant

buffet

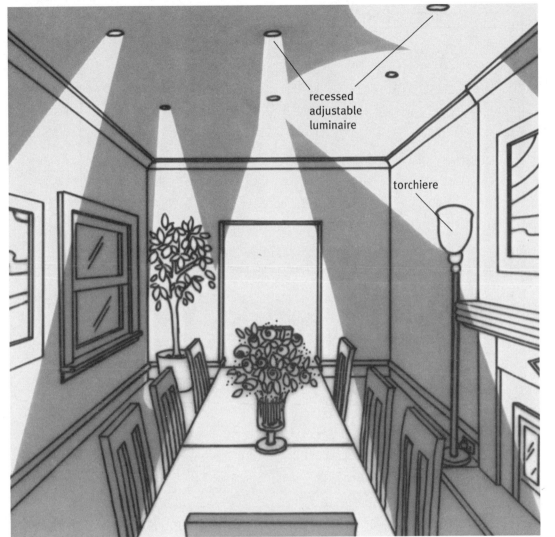

recessed
adjustable
luminaire

torchiere

Drawing 11.13
This perspective drawing
shows what the dining room
looks like with the lighting
and furniture in place (the
floor plan with the lighting
layout is shown on
Drawing 11.12.)

Chapter Twelve

BEDROOMS—PRIVATE SANCTUARIES

This is the one room where ambient light is first and foremost. People are the main event in a bedroom setting. Help erase dark circles and soften age lines. Your clients will love you for it.

Often it's the bedrooms that are thought of as unimportant areas when a lighting plan is being put together. They are often left with a fixture in the center of the ceiling and a couple of half-switched receptacles.

Think about how much time we spend in our bedrooms. Statistics show that we spend one-third of our lives sleeping. Many individuals spend that time sleeping with a significant other(s). People would like to look their best in such an intimate setting.

Ambient Lighting

This is the one room where ambient light is first and foremost. People are the main event in a bedroom setting. Help erase dark circles and soften age lines. Your clients will love you for it.

If there is an existing luminaire in the center of the ceiling, an easy upgrade would be to replace it with a pendant-hung indirect fixture. This will provide illumination that bounces off the ceiling and walls to create a flattering shadowless light (see Drawing 12.1).

Quite often now we see many homes built with sloped ceilings. The tall wall area above the door line is often considered a dead space. Mounting a series of two or three wall sconces up there will create great fill light without wasting any of the art wall space at the normal viewing heights.

Drawing 12.1
Replace the existing ceiling luminaire with a pendant-hung indirect version for a good source of fill light.

Drawing 12.2
A "tray"-style ceiling lends itself perfectly to a cove lighting detail.

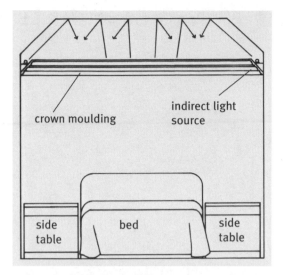

Drawing 12.3
A tall piece of furniture such as this armoire is a great spot to hide a source of ambient light.

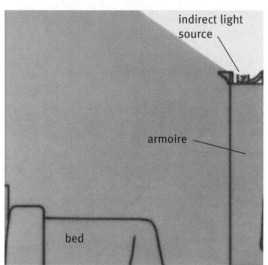

Drawing 12.4
If you choose traditional bedside lamps for reading, specify an opaque liner in the shade, so that the other person in bed can get some sleep.

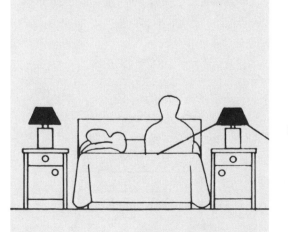

Another architectural phenomenon is the tray or coffered ceiling (see Drawing 12.2). This aspect lends itself to a perimeter cove lighting detail. The light helps emphasize the architect's ceiling concept, while filling the bedroom with ambient light.

Sometimes clients are reticent about putting a lot of time and money into an architecturally-integrated lighting design. For these more cost-conscious customers, a pair of inexpensive torchieres (see Chapter 4, Drawing 4.4) should do the trick.

Also, you could place a halogen indirect light source on top of a tall piece of furniture, such as an armoire (see Drawing 12.3). Often there is a recess on the top of these furniture pieces that provides a great hiding spot for an indirect source of illumination. In some homes there is a canopy bed with a solid top. This, too, could be a great location for a hidden indirect light (see Chapter 4, Drawing 4.8).

Task Lighting

In addition to the ambient light, another function of light should be considered: reading illumination. Typically a pair of portable luminaires (table lamps) are placed on top of the bedside tables. If you choose this approach, there are a few things to consider:

1. Select reading lights that have opaque liners in the shades. This will help direct the light down and across the clients' work surface. Additionally, the opaque shade helps shield the light from the client's bed mate (see Drawing 12.4).

2. Install wall-mounted swing-arm lamps (see Drawing 12.5). They are a flexible source of illumination that doesn't take up space on the bedside tables.

Mounting them at the correct height is critical, however. If they are too high they will be a source of glare. If they are too low, the client may have to slump to an uncomfortable position in order to read. They should be mounted just above shoulder height when sitting in bed.

The best way to find the correct mounting height is to have your client(s) get into

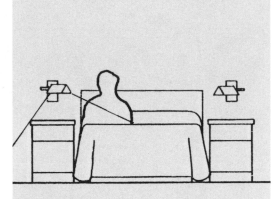

Drawing 12.5
Swing-arm lamps provide good task light, without taking up space on the bedside table. Mount them at just above your clients' shoulder height. Yes, put them in bed and measure.

bed and hunker down against the pillows in their normal reading position. Then measure from the floor to just above their shoulder height. Why? Because the optimum spot to position lighting is between head and work surface. Clients sharing a bed may nest in at different heights. Some compromise should be made on both sides so the reading lights can be mounted at matching heights.

3. A third option for reading lights is to install a pair of recessed adjustable low-voltage luminaires in the ceiling above the bed (see Drawing 12.6).

This is the airline approach to providing light for reading. You may have noticed that the reading lights in the plane are not directly over your seat, but actually over the seat of the person sitting next to you. That's because the airline designers knew that your head makes a better door than a window when it comes to light transmission.

The same principle applies in the bedroom. The client on the right controls the recessed adjustable luminaire over the partner's side of the bed, and vice versa. By using a lamp with a tight beam spread, such as the MR16 ESX (20-watt spot), the light is confined to a circle of illumination about the size of a magazine. If all they read are paperbacks, then you can use a MR16 EZX which projects a very narrow spot.

Using a recessed luminaire that has a small aperture helps reduce the possibility of glare. Some recessed adjustable low-voltage lights come with apertures as small as 1-1/2 inches. It also helps if the luminaire specified has its interior painted black. Otherwise the dichroic reflectors of the MR16 and MR11 create a pink glow of light inside the recessed housing.

Install dimmers on either side of the bed just above the night tables to control the recessed luminaires. Your clients should not have to get out of bed to turn out the lights.

Drawing 12.6 Cross-illumination above the bed provides "quiet" reading light that doesn't disturb the other bed occupant.

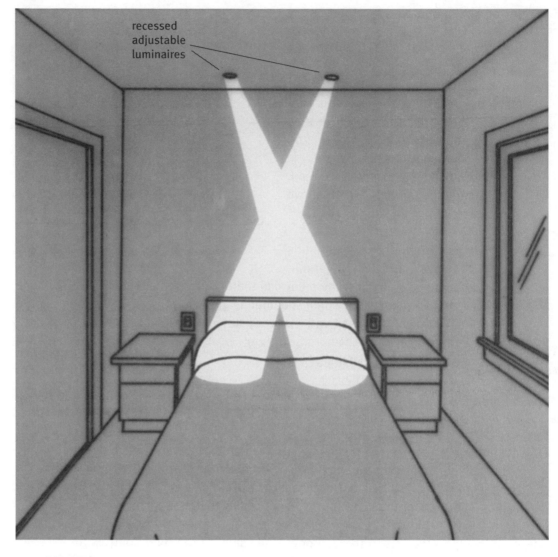

recessed adjustable luminaires

Special Effects

Some designers like to have underlighting (see Drawing 12.7) along the edge of a bed.

This is a more modern effect that creates the illusion of the bed levitating slightly off the floor. It can also be a great night light for those clients who make frequent trips to the bathroom, or occasional trips to the fridge.

Don't use undimmed cooler-colored fluorescent as your source of underlighting. It could make a client look like today's "blue light special." A dimmable fluorescent source in a warm color-temperature would be a more comfortable choice.

Another option is low voltage tube lighting using half watt lamps on 3-inch centers. The tubing is flexible and goes around corners easily. The end effect should be just a gentle glow of light.

Do not laugh, but at least once in your design career someone is going to ask for a mirror above the bed (see Drawing 12.8).

Believe it or not, we often look our best when lying down. Gravity finally works to our advantage, smoothing away lines and excess folds of skin. An old Hollywood trick is for stars to have their close-ups taken while lying on their backs.

So, how do you light people in bed? Cross-illumination, of course. Three recessed adjustable lights flanking the two long sides of the mirror will give a fabulous even illumination. Use MR16's (50-watt FNV) to provide a soft overall glow. If you feel especially accommodating, you can slip in a peach-color filter to add a flattering hue to their skin tone.

One last item to consider doesn't have as much to do with lighting as it does with visibility. Today, new master bedrooms are often quite enormous. The bed will be near one wall and fifteen or twenty feet away on the opposite wall will be an armoire with a television inside. That's simply too far away for people to watch without using binoculars. A more practical solution would be to place a console at the end of the bed that houses a pop-up TV (see Drawing 12.9).

The television disappears into the console when not in use. Think about it. Your clients will be impressed with this great idea.

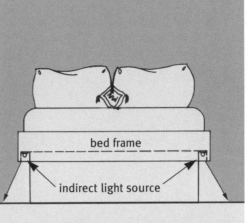

Drawing 12.7
Underlighting the bed gives a sense of hovering, like the bed is about to take off.

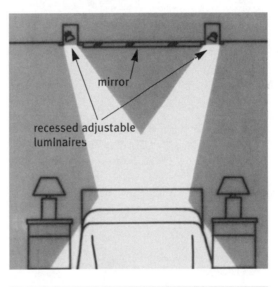

Drawing 12.8
If you have clients who want a mirror above the bed, then cross-illuminate them with three recessed adjustable lights on each of the long sides of the bed, using lamps with complimentary wide beam spreads.

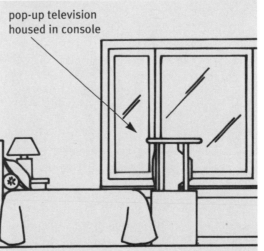

Drawing 12.9
Do your clients a favor and locate the television at the foot of the bed instead of way across the room. There are consoles available that allow the television to pop up when being watched and disappear inside when not in use.

Chapter Thirteen

Bathrooms—Functional Luxury

The most important thing to remember in lighting the bathroom is that good illumination for tasks is primary, because looking good is *hard work*.

Well-designed lighting is of the utmost importance in the bathroom. Yet, more often than not, people use inadequate lighting techniques for much-needed task illumination. How many times have we seen dramatic photograph of a vanity with the recessed downlight directly over the sink? It makes for a great shot, but imagine yourself standing at the mirror with that harsh light hitting the top of your head. Remember when, as a child, you would hold a flashlight under your chin to create a scary face? The same thing happens, only in reverse. Long dark shadows appear under your eyes, nose and chin. This is extremely bad lighting for applying make-up or shaving.

Another typical arrangement is the use of one luminaire, surface-mounted, above the mirror. This is only slightly better than the recessed luminaire. At best, it illuminates the top half of the face, letting the bottom half fall into shadow. This is an especially hard light by which to shave. There are just so many ways to tilt your head to catch the light.

Task Lighting

For the best task lighting, use two luminaires flanking the mirror area above the sink to provide the necessary cross illumination (see Drawing 13.1).

The principle of cross-illumination on the vertical axis originated in the theater, where actors and actresses applied make-up in front of mirrors surrounded by bare lamps in porcelain sockets. In imitation of this technique, luminaire manufacturers about 20 years ago started to put vanity *light bars* on the market. Soon homes everywhere were

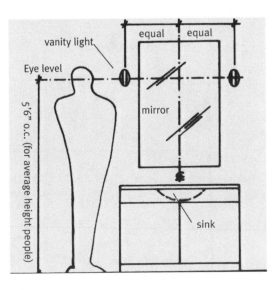

Drawing 13.1
Vanity lights are optimally mounted at eye level, flanking the mirror.

sporting the now ubiquitous three-lamp brass or chrome bar above the mirror. Remember, these bars work best only when mounted on each side of the mirror. A third luminaire could be mounted above the mirror, but it is not necessary for good task lighting. A luminaire mounted above the mirror by itself is not an adequate source of work light.

A more recent trend in providing cross illumination is to wall mount translucent luminaires at eye-level on either side of the sinks (see Drawing 13.2 and 13.3). These can flank a hanging mirror or be mounted on a full wall mirror. For inset sink areas, the mirror lights can be mounted on the return walls (see Drawing 13.4). There are many new well-designed American and European luminaires that are perfect for this type of application. To protect the homeowner from electric shock, luminaires that are located this close to water should be installed with an instantaneous circuit shutoff, called a "ground fault interrupter" (GFI) (see Drawing 13.5).

Drawing 13.2 (left)
Two sinks mounted too far apart need a pair of light luminaires per sink.

Drawing 13.3 (right)
Two sinks mounted closer together can share three light luminaires.

Drawing 13.4 (left)
In small vanity areas, task lights can be mounted on the return walls.

Drawing 13.5 (right)
A GFI (ground fault interrupter) prevents people from being shocked if they touch water and the switch at the same time.

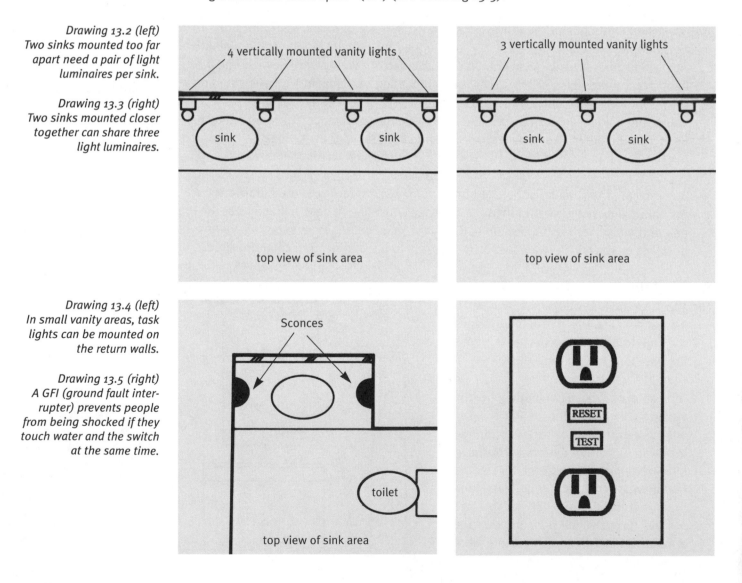

Many builders and architects have a propensity for installing fluorescent or incandescent light in soffits, fitted with either acrylic diffusers or egg-crate louvers, over vanity areas. They, too, mostly illuminate the top half of a person's face. A white or glossy counter can help reflect some light from below by bouncing illumination up onto the lower part of the face. You are cross-lighting from top to bottom in this instance. This is not the optimum solution, but a passable substitute, if vertical cross-illumination is impossible to install. Remember, the more stuff that ends up on the counter, such as towels and containers, the less reflective surface there will be.

While the task area at the vanity is the most critical to illuminate correctly, other areas of the bath bear consideration. Tubs and showers need a good general light. For this purpose, recessed luminaires with white opal diffusers are commonly used and relatively effective. One drawback is that many of the units on the marketplace project approximately two inches below the ceiling line and may not be visually comfortable.

A luminaire with a lens that is flush or recessed into the ceiling might be preferred by those who are sensitive to bright light. However, with such a fully recessed unit, the upper third of the shower or tub area will be a little dimmer (see Drawing 13.6). These luminaires do reduce glare and allow bulbs of higher wattage to be used. Code requirements should be checked when specifying a fixture over a spa or bathtub. The bottom of the luminaire should be seven to seven-and-a-half feet above the high water level.

Make sure that all luminaires in the shower are listed for wet locations by UL, ETL or other approved testing laboratory. If they are tested by Underwriters' Laboratory, they will have a blue UL label. Also check to see if these luminaires should be circuited with GFIs for extra safety.

Fluorescent in the Bathroom

The fluorescent option is important today. Several states demand fluorescent light sources in the construction or remodeling of residential bathrooms because they are at least three times more energy efficient than incandescent lamps. For those designers and architects doing work in California, a state code called Title 24 requires that the general light in bathrooms (and kitchens, as mentioned in Chapter 8) for new construction or remodels of 50% or more must be from fluorescent luminaires (see Drawings 13.7 -13.10).

Fortunately, the color temperature of many of today's fluorescent lamps are very

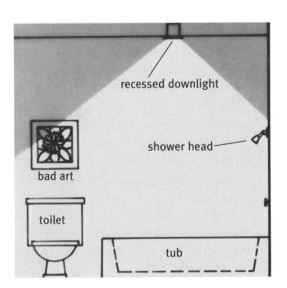

recessed downlight

shower head

bad art

toilet

tub

Make sure that all luminaires in the shower are listed for wet locations by UL, ETL or other approved testing laboratory. If they are tested by Underwriters' Laboratory, they will have a *blue* UL label. Also check to see if these luminaires should be circuited with GFIs for extra safety.

*Drawing 13.6
Recessed downlights may help reduce glare, but the downside is that they aren't adequate sources of fill light.*

Drawing 13.7 (left) California's Title 24 states that if there is only a single light source, then it must be fluorescent.

Drawing 13.8 (right) California's Title 24 states that, in this design, the general illumination must be fluorescent, but the vanity lights can be incandescent or fluorescent.

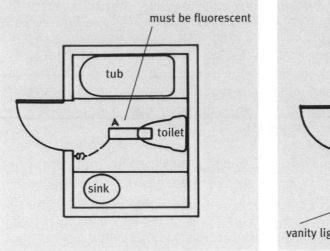

Drawing 13.9 (left) California's Title 24 states that here the part of the bath with the toilet must be fluorescent, while either the general illumination or the vanity lights must be fluorescent.

Drawing 13.10 (right) California's Title 24 states that here, either the vanity lights or the general illumination in the toilet area must be fluorescent.

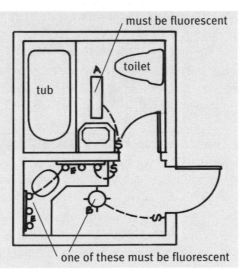

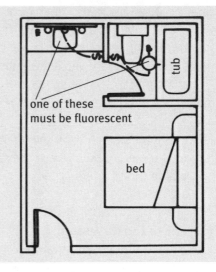

flattering to skin tones. In response to color rendition criticism, most manufacturers have introduced recessed and surface mounted luminaires that use lamps with color-correcting phosphors, including the newer compact fluorescents (CFL's). These lamps not only provide greatly improved color rendering, but the 13 watt version produces an amount of illumination close to a 60-watt incandescent bulb. Many of today's luminaires even use two 13 watt tubes or a 26-watt quad tube that put out as much light as a 120-watt incandescent source for 26 watts of power. Because one of the color temperatures available in the compact fluorescent lamp is close to that of incandescent (2700%K), both light sources can be used in one bath without creating disconcerting color variations.

Windowless interior bathrooms may require an exhaust fan. Specifying a combination fan and light will provide ambient light. Units are now available with a compact fluorescent source, filling energy conservation requirements. Make sure to specify separate switching for the fan and light, if allowed by code.

Two drawbacks to some of the compact fluorescent lamps are an inherent hum and the lack of a rapid-start ballast, the latter deficiency causing the lamp to flicker two or three times before stabilizing. It's a good idea to let clients know the downsides as well as the advantages of this lamp if you specify it. Some quad versions are much quieter and have a relatively rapid start-up. Dimming of compact fluorescents is now a reality. These advances, along with the improved colors, long life, and quiet operation, make fluorescent lighting worth a second look (Chapter 3 covers your fluorescent options).

Ambient Lighting

Indirect lighting in a bathroom adds a warm overall glow to the space. Wall sconces or cove lighting that directs light upward can provide gentle ambient illumination. Both of these can use miniature incandescent lamps, compact fluorescents, or the standard-length fluorescent tubes. The fluorescent choices not only comply with tighter energy restrictions but also provide comfortable low maintenance light for the entire room. For bathrooms with higher ceilings, pendant-hung units can also be considered as a source of fill light.

Skylights

Often skylights are installed to supplement or replace electric lighting during the daytime hours. Clear glass or acrylic skylights project a hard beam of light, shaped like the skylight opening, onto the floor of the bath (as mentioned in Chapter 5). Bronze-colored skylights cast a dimmer version of the same shape, while a white opal acrylic skylight diffuses and softens the natural light, producing a more gentle light that fills the bath more completely. Existing clear or bronze skylights can be fitted with a white acrylic panel at or above the ceiling line to soften the light they cast (see Drawing 13.11).

All specified skylights should have ultraviolet filters to slow the deterioration or fading of materials caused by the sun's ultra-violet rays. If UV filters are not available from the skylight manufacturer, they often can be obtained from companies that manufacture fluorescent outdoor signs. The original fabricator of these filtering sheets of plastic is Rohm and Haas of Philadelphia. Their product is called a UF3 ultraviolet filtering acrylic sheet.

If the light well is deep enough, low-maintenance fluorescent strip lights or luminaires with long-life incandescent lamps can be mounted between the acrylic panel and the skylight(see Drawing 13.11). These inexpensive lights can be used to keep the skylight from appearing as a dark recess in the ceiling at night.

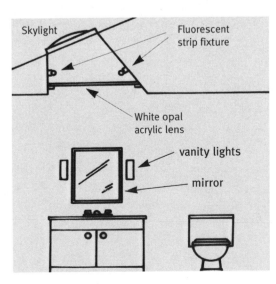

Drawing 13.11
Lighting can be located within the light well to provide additional illumination at night, instead of becoming a "black hole" at night.

If fluorescent luminaires are used, dimming ballasts can be specified to allow control over the amount of light being emitted. Although such ballasts are not inexpensive, one of the advantages of dimming fluorescents is that they don't change color temperature when dimmed. In this regard, they are unlike incandescents, which become more amber as they are dimmed.

Another good reason for adding some type of ambient illumination in bathrooms is that they are becoming multi-functional areas. Homeowners now have exercise areas, dressing rooms, lounging areas and whirlpools for more than one. Some bathrooms have become intimate entertaining areas that deserve all the design care you give to the other main areas in the house.

Accent Lighting

Along with this new-found need for ambient illumination comes an opportunity for accent lighting. Plants and art pieces can be highlighted.

When clients are entertaining, the room most frequently visited by their guests will likely be the powder room. This space can be treated differently than the other bathrooms. No serious tasks are going to be performed by guests.

This is a place where people will wash their hands or check their hair and make-up before rejoining the soirée. The lights should be just a flattering glow. Sometimes a pair of translucent luminaires on either side of the mirror or a single wall sconce will do the trick.

Some powder rooms do double duty as guest baths for overnight house guests. If this is the case, light the bath as you would a master bath, but make sure to put the various lights on dimmers to allow for flexible control over the illumination levels.

The most important thing to remember in lighting the bathroom is that good illumination for tasks is primary, because looking good is *hard work*.

Chapter 14

HOME OFFICES—WORK SPACES THAT REALLY WORK

Designers seek a more residential feel for work environments. They gravitate toward color, texture, round edges, plush carpeting, plants, and warm comfortable illumination. This design trend has made its mark as America's work force is moving back home.

With the advent of the personal computer and the fax machine, an amazing number of people have moved their offices into their homes. Gone is the aggravation of the commute, along with the overhead of a separate work space. Some designers and related professionals find that they spend most of their time with clients at job sites or showrooms, making an office, as a separate entity, unnecessary.

For many more individuals the home office serves as an important supplement to the main office space. Information can be input through the home personal computer, faxes can be received and sent, and a few productive hours of work can be put in without leaving the house.

The question is, how do you make this office space usable and comfortable for yourself or for your client?

Often the office area is visible from the rest of the house. How do you make it truly work-oriented without creating a commercial looking office environment? The trend to humanize office environments has been going on for the last seven or eight years. The early to mid-eighties were the culmination of years of hard-edged commercial design.

Lighting design seemed to follow the same route. The trend was to fill rooms with flat, shadowless, almost hospital-like illumination. Depth and dimension were lost. Softness and texture were gone.

Finally, people grew tired of it. Now, designers seek a more residential feel for work environments. They gravitate toward color, texture, round edges, plush carpeting, plants, and warm comfortable illumination. This design trend has made its mark as America's work force is moving back home.

A common misconception is that the more light there is, the better people can see.

Ambient Lighting

With business now so computer-oriented, a designers' major focus shifts to how lighting affects people working at a computer's VDT (video display terminal). Since much of our work these days is done on the computer, we have genuine concerns about eye strain, fatigue and headaches. These can often be caused by contrast and reflection. Glare is caused by contrast and reflection.

Surprisingly, even what you wear will affect the reflection. A white shirt or blouse will show up on the screen and keep you from seeing areas of the data display (see Drawing 14.1).

Here are some considerations when putting together a design:

1. A common misconception is that the more light there is, the better people can see.

2. In the case of VDTs, if there is too much light in the room, the screen becomes difficult to read (see Drawing 14.2).

3. Another factor is the difference between the surface brightness of the document from which the computer operator is taking information and the brightness of the screen itself. Going back and forth from a brightly illuminated document to a more dimly lit screen will cause eye fatigue.

4. The fourth problem is reflections on the screen. A ceiling filled with recessed luminaires or track light can cause little glare spots on the mirror-like surface of the screen (see Drawing 14.3).

What steps can you take to minimize these problems? A lot of the answer relates to the interior design and space planning.

1. Step one would be to keep the color contrast between the walls and ceiling to a minimum. Stay away from using glossy finishes that could reflect onto the screen.

2. Another planning consideration is to avoid positioning people with their back to a window. The image of a bright window will overpower data on the VDT.

3. There are anti-glare screens available, but you can't rely on them totally. They must be used in conjunction with proper lighting.

4. If recessed or track lighting is being considered, then specify units that have louvers or baffles that create a virtual cut-off at the face of the unit. Filters have also been used as a way of reducing glare for VDT screens situated at certain angles to the luminaires.

Instead of using recessed luminaires or track lighting, consider using a source of ambient illumination instead. Torchieres would be a serviceable solution in a home office setting

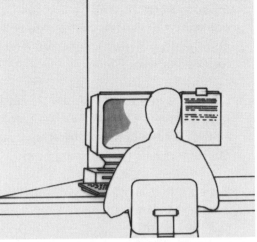

Drawing 14.1 (left)
Even a light-colored shirt or blouse can cause a reflection on the screen, obscuring some of the data displayed.

Drawing 14.2 (right)
Positioning the screen away from the windows or other highly reflective surfaces will help reduce glare on the screen.

Drawing 14.3
A series of recessed luminaires will cause hot spots of glare to appear on the VDT.

*Drawing 14.4
Placing an uplight
behind the computer can
help lessen the contrast
between the screen and
the wall color.*

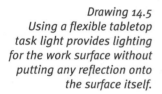

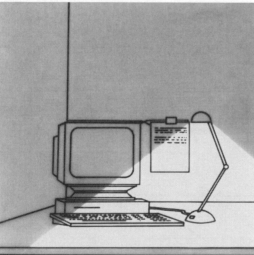

*Drawing 14.5
Using a flexible tabletop
task light provides lighting
for the work surface without
putting any reflection onto
the surface itself.*

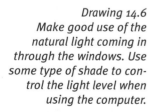

*Drawing 14.6
Make good use of the
natural light coming in
through the windows. Use
some type of shade to control the light level when
using the computer.*

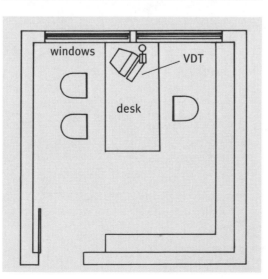

(see Chapter 4, Drawing 4.4). They can produce a soft, relatively even illumination across the ceiling and screen without the typical hot spots created by recessed or track lighting.

5. If the VDT is in a corner, an uplight positioned behind the screen will help soften the contrast between the screen and the corner walls (see Drawing 14.4).

6. Make sure that light coming through the windows is controllable, using some sort of shade or blind (see Drawing 14.6).

Task Lighting

Layered with this general illumination should be some flexible task-oriented lighting. A tabletop or wall-mounted luminaire that has a flexible arm and some variability in light levels would be a good way of lighting documents and the keyboard without spilling light onto the screen itself. Select a luminaire with a opaque shade, so that the image of the luminaire itself won't reflect onto the VDT (see Drawing 14.5).

Daylight

Remember that natural light can be a very usable source of illumination and should be utilized to the best advantage (see Drawing 14.6).

Chapter Fifteen

EXTERIOR ROOMS—EXPANDING INTERIOR SPACES VISUALLY

In reality, the illumination of exterior spaces can be directly related to how the interior areas are perceived. One of the great benefits of exterior lighting is that it can visually expand the interior rooms of a residence.

As a designer of interior spaces, you may not feel a need to incorporate exterior lighting as a part of the overall design. In reality, the illumination of exterior spaces can be directly related to how the interior areas are perceived. One of the great benefits of exterior lighting is that it can visually expand the interior rooms of a residence. When there is no illumination outside, windows become highly reflective at night. This is known as the "black hole" effect. The windows end up reflecting the lights in the room, so that all the clients can see at night is their own reflection instead of the view beyond (see Drawing 15.1).

Unfortunately window treatments only tend to accentuate the problem by "framing" the black hole. Closing the draperies only creates the illusion of a smaller size room.

People often feel boxed-in at night when they are surrounded by these "black holes". The rooms can seem smaller than they actually are. The rule of thumb is to try and balance the amount of light inside and outside the house. This allows the windows to become more transparent, as they are during the day.

Psychologically, too, people feel safer when they can see the yard area around them. They feel more vulnerable inside the house when there is no lighting outside.

You don't have to light up the exterior like the White House. That type of illumination would come under the heading of security lighting.

Security Lighting
Security lighting and landscape lighting are

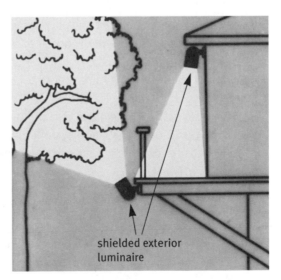

shielded exterior luminaire

Drawing 15.1
Fore lighting the deck from under the eaves and uplighting the trees from under the deck will help make the outside feel like part of the interior spaces.

Some areas have "light pollution" regulations requiring fixtures to be shielded or aimed to prevent illumination that may disturb the neighbors.

two different things. Yet, often you will see people trying to use the same lights to perform both functions.

Security lighting is often simply a source of lighting that immediately floods the yard with a good punch of illumination. This is what clients turn on when they hear a noise outside. Security lighting does not have to be so confrontational, the main objective is to provide enough light so that homeowners can see what is causing the disturbance outside.

Security lights should be mounted as high as possible under the eaves. Specifying "double bullets" (see Drawing 15.2) aimed in different directions and angled out and down will provide good yard illumination.

Some areas have "light pollution" regulations requiring fixtures to be shielded or aimed to prevent illumination that may disturb the neighbors (see Drawing 15.3).

Drawing 15.2 (left) Security lighting is normally mounted at the roof-line along the perimeter of the house.

Drawing 15.3 (right) Reasonably priced bullet shaped luminaires are a good source of security lighting.

Security lighting is optimally controlled by a *panic switch* located next to the bed in the master bedroom and in the bedroom of another responsible person in the household, such as the grandmother, au pair, or oldest child. These security lights can also be controlled by a motion sensor (see Chapter 6).

Security lights do not come on as part of the landscape lighting. There is nothing worse than driving up to someone's home, only to be assaulted by glaring security lights mounted on the corner of the house. As a guest you may feel like you've been caught in the middle of a prison break.

Landscape Lighting

What do you do when you have a client who's installed this kind of system, thinking they've got landscape lighting? One approach would be to say, "Oh, how wonderful; you've put in an effective security lighting system. Now let's talk about the landscape lighting." This way they don't lose face, and are more open to ideas on additional lighting concepts.

Landscape lighting needs to be subtle. Attention should be drawn to the plantings, sculpture and out-buildings, not the luminaires.

Decorative exterior luminaires such as lanterns can't do the job by themselves. They can easily overpower the facade of the house and the yard area if they are the only source of illumination.

Typically you will see two lanterns flanking the front door and maybe a post light at the end of the driveway. These just become disturbing hot spots that leave everything else in silhouette (see Drawing 15.4).

Still, they play an important role in the overall lighting design. Their job is to create the illusion that they are providing all of the exterior lighting, when, in reality, they should have no more than 25-watts worth of illumination. They serve the same purpose as interior decorative luminaires.

Another aspect to consider when selecting an exterior lantern is the glass. Too often, they are chosen with a clear or beveled glass. The result is that at night people only see the lamps inside, instead of the luminaire itself. If you choose a luminaire that has a frosted glass, an iridescent stained glass, or a sandblasted seedy glass, then the volume of the lantern is seen instead of just the light bulb (see Drawing 15.5).

If the lanterns are existing, it is possible to have the glass in them sandblasted. Often, mirror companies also do sandblasting as a side line. Remember to have only the inside sandblasted. If you do the outside, fingerprints will show because of the oil in our skin.

Correct sizing of exterior fixtures can be tricky. Lanterns displayed in lighting showrooms

If the lanterns are existing, it is possible to have the glass in them sandblasted. This helps obscure the view of the light bulb within.

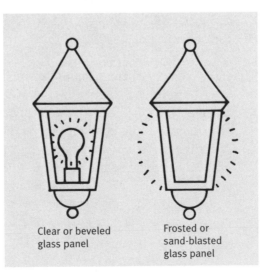

Clear or beveled glass panel   Frosted or sand-blasted glass panel

Drawing 15.4 (left) Decorative luminaires alone become too predominate, leaving the rest of the yard and house in relative darkness.

Drawing 15.5 (right) Selecting a luminaire that has clear or beveled glass shows off only the light bulb at night. Specifying sandblasted or frosted panels allows the luminaire itself to be the focal point.

Voltage drop— a loss of electrical current due to overload or long length runs (usually over 100 feet), causing lamps at the end of the run to produce a dimmer light than those at the beginning of the run.

appear about 25% larger than they do when installed on a home. The eye tends to make a visual room out of the surrounding fixtures so the lantern is viewed in a very small space.

Cut out a piece of cardboard the size of the prospective lantern. Hang it on the house and then back away. View it from the street or driveway.

Landscape Lighting Techniques

In designing the landscape lighting a decision must be made as to which voltage system will be used, line voltage or low voltage.

If the landscaping has already been done and you are adding the lighting installation of a 120 volt system, it might be expensive and disruptive to the plants due to trenching. 120 volt systems require the wiring to be buried in conduit or the use of direct buried wire. Luminaires for 120 volt systems are often larger than those using 12 volt lamps.

Check local codes for permit requirements. 12 volt lighting systems are less restrictive and installation is relatively easy. The low-voltage 12 gauge cable does not have to be buried, but hiding it under a layer of bark or a shallow layer of dirt is more visually appealing.

Low-voltage systems can use much less power and may not require any additional circuits.

The flexibility of a low-voltage system makes altering the original lighting design feasible and easy without costly rewiring.

When laying out a 12 volt system remember to include the transformer. Voltage drop must be taken into consideration if it is necessary to a have run of over 100 feet.

Voltage drop - a loss of electrical current due to overload or long length runs, (usually over 100 feet) causing lamps at the end of the run to produce a dimmer light than those at the beginning of the run.

There are many techniques for landscape lighting from which you can choose. When working with new construction it is important to specify a number of outside duplex GFI receptacles for future landscaping or portable luminaires for parties. Planning ahead for switching and transformer locations will save your clients money when the landscaping is started. Save time and money by having power lines or conduit installed under the driveway or patio before paving or bricking. Small plants and trees grow, some slowly and some rapidly. Plan for maximum growth and install smaller wattage lamps that can be replaced with higher wattages as the foliage matures. Using a variety will keep the design interesting. Using only one technique may create a too commercial-looking design. Here are some options for you to consider:

Uplighting—This can be a very dramatic way of lighting trees that have a sculptural quality to them (see Drawings 15.6 and 15.7).

The luminaires can be ground-mounted or actually installed below-grade. These buried luminaires are known as *well lights*. Well lights have little or no adjustability, so they work best for mature trees.

Above-ground directional luminaires have a much greater flexibility and therefore do a better job for younger trees as they mature. Use shrubbery to conceal the light source from view. A below-grade junction box will allow the luminaire to be closer to ground level.

Silhouetting or Back Lighting—There are now fluorescent luminaires that do a good job of wall-washing, consuming a small amount of power with a long lamp life (see Drawing 15.8). Remember to specify a ballast designed for low temperatures if your project is located in a cold part of the country.

Downlighting—This type of lighting is to be used for outdoor activity areas. It's best to overlap the spreads of illumination to help reduce shadowing. The luminaires can be mounted on trellises, eaves, gazebos and mature trees (see Drawing 15.9).

Spot Lighting—Use this technique minimally. Statues, sculpture or specimen plants deserve to be highlighted. They will tend to dominate the view as people look outside (see Drawing 15.10). Spot lights should be shielded to avoid glare if they are in direct view.

Drawing 15.6
When the lighted object can be viewed from one direction only, above-grade accent lights are the logical choice. To prevent direct glare, fixtures are aimed away from observers. Place the accent lights behind shrubbery to keep a natural looking landscape.

Used by permission of Kim Lighting

Drawing 15.7
If the lighted object may be viewed from any direction, well lights are the ideal solution. These below-grade luminaires are louvered to further reduce the potential for glare. Use the optional directional louver to gain efficiency when the lamp must be tilted within the well.

Used by permission of Kim Lighting

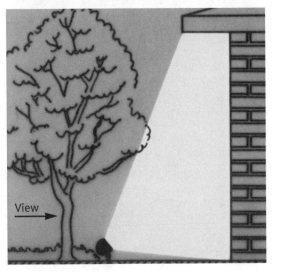

Drawing 15.8
Trees and shrubs with interesting branch structure are dramatic when silhouetted against a wall or building facade. This combination of landscape and facade lighting provides additional security near the building.

Used by permission of Kim Lighting

*Drawing 15.9
Down-lighting — For outdoor
activity areas, luminaires
placed above eye level provide efficient lighting for
recreation, safety and security. Overlapping light patterns will soften shadows
and create a more uniform
lighting effect. Mount to
trellises, gazebos, facades,
eaves or trees.*

*Used by permission of
Kim Lighting*

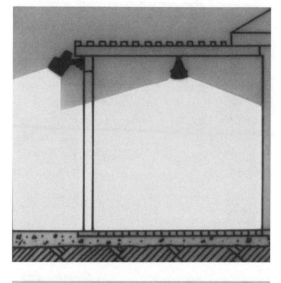

*Drawing 15.10
Spotlighting — Special
objects such as statues,
sculpture, or specimen
shrubs should be lighted
with luminaires that provide
good shielding of the lamp.
Mounting lights overhead
on eaves or trellises helps
reduce glare. If ground-
mounted luminaires are
used, conceal them
with shrubbery.*

*Used by permission of
Kim Lighting*

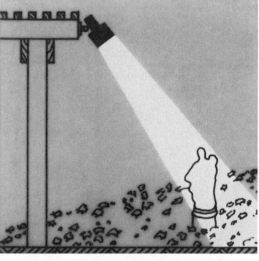

*Drawing 15.11
Fully shielded mushroom-
type lights highlight pathways and ground cover
without drawing attention
to themselves.*

*Used by permission of
Kim Lighting*

fully shielded light source

Path Lighting — This is one lighting technique that needs to be done judiciously. Too
often we see walkways or driveways flanked
with rows of pagoda lights as the only
source of exterior illumination. This tends to
look like an airport runway.

When a pathway light is needed, consider
using an opaque mushroom-type luminaire
that projects light down without drawing
attention to itself (see Drawings 15.11 and
15.12). The luminaires should not exceed 14
inches in height. This, in combination with
additional lighting sources will help create a
comfortable exterior environment.

Spacing of path lights will depend on the
style of the luminaire and lamp options.
Many lighting showrooms now have landscape displays to help you make an
informed choice.

Step or Stair Lighting — Fixtures can be
recessed in the side walls or the steps themselves to illuminate the risers. This will provide safety as well as background fill illumination for the landscape design (see
Drawing 15.13)

Moonlighting — This is the most naturalistic
way of lighting an exterior space. The effect
is as if the area were being illuminated by a
full moon. A dappled pattern of light and
shadow is created along pathways and over
low-level plantings.

This is accomplished by mounting luminaires in mature trees, some pointed down
to create the patterned effect and some
pointed up to highlight the foliage canopy
(see Drawing 15.14)

Controls — It's best not to dim exterior lighting. Many outdoor luminaires use incandes-

cent sources. When incandescent lamps are dimmed, the light becomes more amber. The yellow cast makes the plantings look sickly. The whiter the light, the more healthy the plants look. You can divide the lights into different switching groups. A typical arrangement would be to have the decorative exterior lights on one switching group, possibly on a timer that would come on and go off even if your clients aren't home. The second group could be the accent lighting throughout the yard, and the third would be the security lighting.

Please note that GFI's (ground-fault interrupters) are required on all outdoor circuits in some areas. Local codes should be checked prior to installing the system.

If your client wants to dim the outside lighting, it is possible for the GFI's to be tripped erroneously if the lights are dimmed with phase-controlled dimmers.

Filters—It's best to stay away from colored filters. They tend to change the look of the plantings to an unrealistic color.

When designing with incandescent sources, one filter you can consider is what's called a "daylight-blue filter". It filters out the amber hue of incandescent light to produce a blue-white light that is very complimentary to plants, making them look lush and green. Many manufacturers offer daylight-blue filters as an option. Sometimes they are called "ice-blue" or "color correction" filters. This small addition can make a huge difference in the overall look of the landscape lighting. Refer back to Chapter 2's section on color temperature and plants, and Chapter 4's section on color filters.

partially shielded light source

Drawing 15.12
Partially-shielded path lights create a useful effect when placed within taller shrubs. The surrounding shrubbery filters light onto the pathway, reducing glare, while also lighting the adjacent shrubs.

Used by permission of Kim Lighting

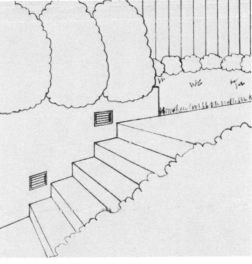

Drawing 15.13
Step lights can be mounted in the side walls of the stairway to provide safe illumination.

Drawing 15.14
Moonlighting—The effect of moonlight filtering through trees is another pleasing and functional outdoor lighting technique. Both up and down lighting is used to create this effect. With luminaires properly placed in trees, both the trees and ground are beautifully illuminated. Ground lighting provides security, and is accented by shadows from leaves and branches.

Used by permission of Kim Lighting

Please note that GFI's (ground-fault interrupters) are required on all outdoor circuits in some areas. Local codes should be checked prior to installing the system.

There are many more exterior luminaires using fluorescent and H.I.D. sources on the market suitable for residential installation. Mercury vapor and metal halide, as well as the cooler-colored fluorescent sources, can do a wonderful job of providing a crisp blue-white light.

Chapter Sixteen
APPLYING LIGHTING TECHNIQUES

This section allows you to hone your lighting design skills. It is a chance to combine what you already know with the information and ideas put forth in this book.

Part A shows guidelines for drawing a reflected ceiling plan

Part B provides guidelines for writing lighting specifications in a clear and complete manner.

Part C is a questionnaire form that you can give to clients in order to get them thinking about lighting and the various functions it serves. The more they understand the importance of well-integrated lighting, the more they may be willing to invest.

Part D is a sample lighting project on which you will provide the lighting layout for a condominium project. A sample layout is provided in Appendix One.

Part E is a review exam covering much of the information covered in Chapters 1 through 15. Sample answers are provided in Appendix One.

Part A—Guidelines for Drawing a Reflected Ceiling Plan (Lighting Plan)

Everyone has different styles of presentation. The important thing is to get information across clearly and concisely. Remember, your lighting plan must be read and understood by people who aren't familiar with your design: architects, interior designers, contractors, and clients.

Start with the plan of the building (or room), with all walls and door swings as background. Add to it (in dashed lines) all ceiling features, overhangs, skylights, columns, and beams that have a bearing on your lighting design.

Draw your luminaires using clear and consistent lighting symbols. Be sure to include a symbol list (legend) on the drawing. If you have any unusual luminaires, try to use a simple symbol which suggest that luminaire. This simplicity saves drawing time and is less confusing or distracting to the reader of the drawings. (see Chart 16.2)

On your plans, be sure to provide dimensions, distances, and spacings of your luminaires with respect to the architectural elements, because the location of the luminaire is critical to your design. In most cases, lighting symbols are larger than the actual luminaires you specify, and the drawing may not be accurately drafted for a number of reasons. As a result, it may not be safe to allow the drawing users to scale off your drawing for luminaire locations. Show dimensions to be sure! (see Chart 16.1)

Drawing 16.1
Make your lighting plan as clear as possible. The more information you provide, the less time you will spend on the phone or at the site explaining what you want done.

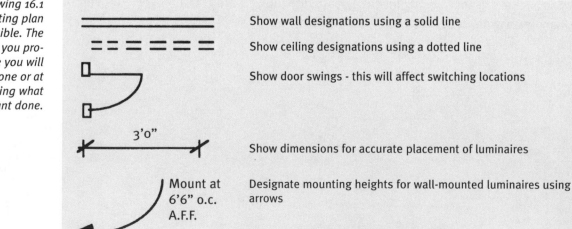

Show wall designations using a solid line

Show ceiling designations using a dotted line

Show door swings - this will affect switching locations

3'0"

Show dimensions for accurate placement of luminaires

Mount at 6'6" o.c. A.F.F.

Designate mounting heights for wall-mounted luminaires using arrows

Part B—Guidelines for Writing Lighting Specifications

A specification (or "spec") is a detailed description of the lighting equipment you have chosen for your project. This is a document which will be used by contractors, distributors, and manufacturers who want to bid on the project, so you must take care to indicate exactly what you want, or else risk getting inferior luminaires on the job, luminaires with finishes that don't blend with the room, luminaires that look awkward or inappropriate, and luminaires that don't distribute the light in the way you intended or outlets where they don't belong.

Cover the following factors in your legend or on your plans:

1. Dimensions of luminaire (such as 2' X 4' recessed fluorescent)

2. Orientation of luminaire (recessed, surface, wall-mounted)

3. Finish of luminaire (white, chrome, gloss, matte, etc.)

4. Luminaire mounting height or suspension length

5. Type of lens or louver or baffle, if appropriate

6. Luminaire light distribution characteristics (e.g. asymmetric forward throw) if there is an option on the luminaire

7. Number of lamps (bulbs) per luminaire, lamp types, and color temperature(Degrees Kelvin, for fluorescent or H.I.D. sources only), beam spreads and voltages.

8. Ballast characteristics, such as dimming capabilities and voltage

Set-up for legend:

1. Generic description of luminaire or component

2. Manufacturer's name and catalogue number

3. Finish

4. Lamping (plus louver or other attachments if needed).

5. Voltage

6. Special notes, i.e., contact person for custom designs

Examples:

wall sconce, Halo H2571 white, with Q150T3CL, 120V.

exterior wall-mounted directional luminaire, Hubbell 309-14 (black) with 75HIR38 flood with #1338 louver, 120V.

recessed adjustable low-voltage integral transformer luminaire, Lightolier 1102 P1 1152 white, with 50-watt MR16 EXN, 12V.

line voltage dimmer, Lutron Diva DV600P white

You will notice in the sample legend (Chart 16.2) that a half-moon symbol always represents a wall sconce. If you have more than one style, designate each type with a letter: For example, if a symbol represents a wall sconce, and you have more than one style, designate each type with a letter, such as Ⓐ, Ⓑ , etc.

Recessed luminaires in the sample legend are represented by circles within squares. For example, a ⊡ represents a recessed adjustable low-voltage luminaire, while a ⊡ represents a line-voltage downlight. Adding an "S" in the circle indicates that this downlight is wet-location rated for a shower or tub area.

These symbols, along with the names of the manufacturers you use most often can be input into your computer to generate clear explicit legends for each of the projects that include lighting.

Building a data base will save time on future jobs. The legend can then be printed on a transparent matte sticky-back sheet, which can be affixed directly to the drawing if your office is not yet on a CAD system (computer-assisted drafting).

Specification Book—Assemble a specification book that has photocopies of all the fixtures and related components such as the dimmers that you recommend. This will further clarify to the clients and contractors what you intend to have installed. A clear picture will answer a lot of questions in advance.

It's also a good idea to draw the corresponding symbols at the top right-hand corner of each specification sheet to match the legend to the actual luminaire being specified. See samples 16.3 - 16.4.

Note that the manufacturer's name is circled, along with the catalogue number and any other pertinent information. The upper right-hand corner shows the corresponding symbol.

Recessed Fixtures

Recessed Fluorescent downlight., Lightolier 1102-FL-1128 with one F13DTT/27K, 120V

Recessed low-voltage mirror reflector fixture with integral transformer, Capri LV1EX/120-LC105 with 50 watt MR16EXn,(Sylvania), 12V

Recessed low-voltage adjustable fixture with integral transformer, Lightolier 1000LV-1052LV with 50-watt MR16EXT, 12V

Recessed low-voltage adjustable fixture with integral transformer for insulated ceiling, Lightolier 1000ICV-1052LV with:
S = Spot Bulb, 35 Watt MR16FRA, 12V
F = Narrow flood bulb, 35 watt MR16FMW, 12V

Existing recessed fixture to be retrimmed with Halo 1450P with 50 watt MR16EXN, 12V

Recessed low-voltage adjustable fixture with integral electronic transformer for insulated ceiling, Juno AL44-445 with 50 watt MR16EXN, 12V (or equal)

Recessed downlight for insulated ceiling, Lightolier 1100IC-1128 with 60 watt A19IF, 120V

Recessed watertight incandescent downlight, Lightolier 1100IC-1128T (wet location) with 60A19IF, 120V

Low-voltage well light, Hadco IL116G-H with 50 watt MR16EXN, 12V (location of transformer to be determined by electrical contractor)

Ceiling Mount - Strip/Skylight

Fluorescent strip fixture for closets, Wellmade 183-A-224TS-HPF with two F20T12CWX, 120V

Surface-mounted fixture, such as Prisma 1520 (Delta Tondo), with 75-watt A19IF, 120V

Chart 16.2
This is a sample legend, showing the symbol, generic description and catalogue number for each luminaire. Keeping this information on computer will speed up the generation of future legends.

Recessed Fixtures - continued

Surface-mounted fluorescent 2'x2' fixture, Forcast Lighting 2922-75-F22 with two F40U/4K, 120V

Fluorescent strip fixture for closets, Wellmade 183-A-224TS-HPF with two F20T12CWX, 120V

Fluorescent under-cabinet task light, Alkco SFHP213 (prismatic diffuser), 42-1/2", with two 13-watt T5/4K, 120V

Alkco CLI2000 series (dark bronze, other finishes available), direct wire, with 25-watt T6-1/2IF at 8" o.c. spacing, 120V
Both standard and custom lengths are available. Contractor to field measure for exact lengths needed.

Existing surface mounted fluorescent fixture to be removed.

Strip/Track Lights

Miniature low-voltage track light system, Track=Lightolier basic Lytespan 1-circuit track, Lightolier 6002 BK black. The track should be a continuous run along the side of the beam. (Contractor to field measure for exact lengths needed). Fixture= Lightolier Prevue 6416 (Black) track heads with 50-watt MR16 EXN, 12V

Miniature Low Voltage strip fixture, Low-voltage strip fixture, Starfire Xenflex SF-3-5-24-XF-RF (WHite reflector), 24V., Contractor to field-measure for exact lengths needed. (Remote transformer to be specified by electrical contractor.)

Low-voltage directional SF12V "halogen bridge" system (custom lengths to be field-measured at site) with Byrdy (aluminum) with 50-watt MR16EXN-silver reflector, 12V., with remote transformer by SF12V.

Vanity Fixtures

Wall-recessed vanity fixture with integral transformer, Zelco ZLW0234, with 20-watt halogen, 12V., Final choice by owners/interior designer.

Vanity fixture sucha as Georgian Art Miranda 20-1951 - rust with one 60-watt candelabra, 120V. Final choice by owners/interior designer.

Wall bracket vanity light, such as Halo H2592 with 100A191F, 120V. Final choice by owners/interior designer.

Mirror light, such as French Reflection Miroir Brot duplex long-arm with extra-strength magnification glass (3x), available through Casella Lighting, 111 Rhode Island, San Francisco, 415 626-9600.

Vanity fixture, such as IPI Lighting Kumo II with 75-watt A19IF, 120V. Final choice by owners/interior designer.

Wall Sconces

Wall sconce, such as Ron Rezek 471 with one 100A19IF, 120V. Final choice by owners/interior designer

Wall sconce, such as Phoenix Day 3630 with two 75-watt A19IF, 120V. Final choice by owners/interior designer

Wall sconce, such as Justice Design Group 4000 (Crest) with one 150-watt A19IF, 120V. Final choice/color by owners/interior designer

Fluorescent wall sconce, such as Ron Rezek Scroll with two PL13 Quartz/27K, 120V. Final choice by owners/ interior designer

Pendant-Hung/Chandeliers

Pendant-hung fixture, such as Floss Trama Suspension 25" with counterweight, with A300T3CL, 120V. Final choice by owners/interior designer.

Chandelier, such as Lightspann Cuckoo-nest pendant. Final choice by owners/architect. Available through Studio One, 415 861-7200, Space 310, Galleria Design Center, 101 Henry Adams St., San Francisco

Low-voltage pendant-hung fixture, such as Kock & Lowry C-330 Vodka fixture with 20-watt halogen, 12V. Final choice by owners/architect.

Existing chandelier to be removed.

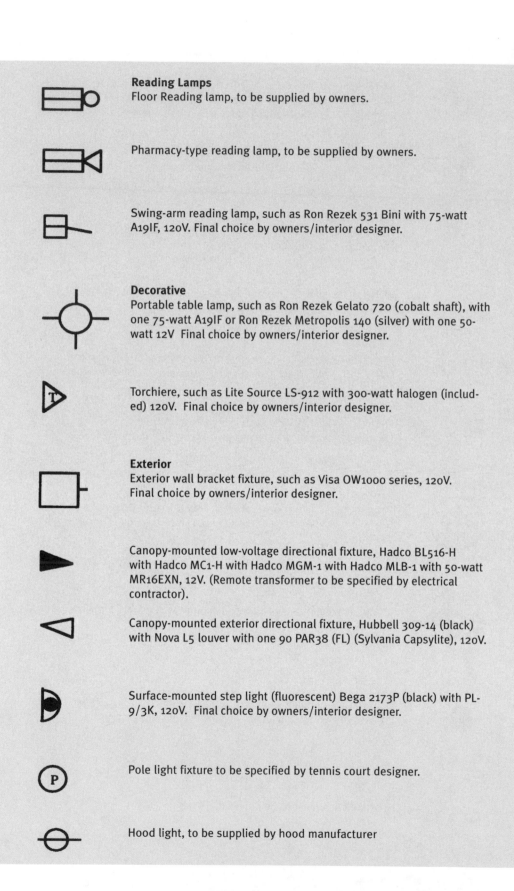

Reading Lamps
Floor Reading lamp, to be supplied by owners.

Pharmacy-type reading lamp, to be supplied by owners.

Swing-arm reading lamp, such as Ron Rezek 531 Bini with 75-watt A19IF, 120V. Final choice by owners/interior designer.

Decorative
Portable table lamp, such as Ron Rezek Gelato 720 (cobalt shaft), with one 75-watt A19IF or Ron Rezek Metropolis 140 (silver) with one 50-watt 12V Final choice by owners/interior designer.

Torchiere, such as Lite Source LS-912 with 300-watt halogen (included) 120V. Final choice by owners/interior designer.

Exterior
Exterior wall bracket fixture, such as Visa OW1000 series, 120V. Final choice by owners/interior designer.

Canopy-mounted low-voltage directional fixture, Hadco BL516-H with Hadco MC1-H with Hadco MGM-1 with Hadco MLB-1 with 50-watt MR16EXN, 12V. (Remote transformer to be specified by electrical contractor).

Canopy-mounted exterior directional fixture, Hubbell 309-14 (black) with Nova L5 louver with one 90 PAR38 (FL) (Sylvania Capsylite), 120V.

Surface-mounted step light (fluorescent) Bega 2173P (black) with PL-9/3K, 120V. Final choice by owners/interior designer.

Pole light fixture to be specified by tennis court designer.

Hood light, to be supplied by hood manufacturer

Half-switched receptacle.

Switched floor receptacle.

Floor receptacle.

MCS   Momentary Contact Switch (door jam switch).

Junction box, to be stubbed for future landscape lighting.

Ceiling fan/light, such as Casablanca Aritian with integrated halogen fixture, 120V. Final Choice by owners.

—3 S —   3-way switch.

—4 S —   4-way switch.

©—   Fan/light control, to be specified by electrical contractor.

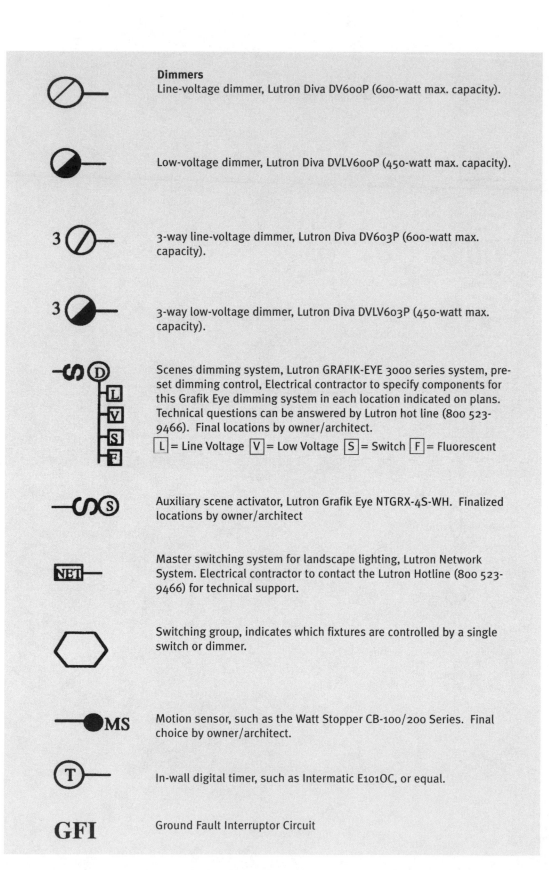

Dimmers
Line-voltage dimmer, Lutron Diva DV600P (600-watt max. capacity).

Low-voltage dimmer, Lutron Diva DVLV600P (450-watt max. capacity).

3-way line-voltage dimmer, Lutron Diva DV603P (600-watt max. capacity).

3-way low-voltage dimmer, Lutron Diva DVLV603P (450-watt max. capacity).

Scenes dimming system, Lutron GRAFIK-EYE 3000 series system, pre-set dimming control, Electrical contractor to specify components for this Grafik Eye dimming system in each location indicated on plans. Technical questions can be answered by Lutron hot line (800 523-9466). Final locations by owner/architect.
L = Line Voltage   V = Low Voltage   S = Switch   F = Fluorescent

Auxiliary scene activator, Lutron Grafik Eye NTGRX-4S-WH. Finalized locations by owner/architect

Master switching system for landscape lighting, Lutron Network System. Electrical contractor to contact the Lutron Hotline (800 523-9466) for technical support.

Switching group, indicates which fixtures are controlled by a single switch or dimmer.

Motion sensor, such as the Watt Stopper CB-100/200 Series. Final choice by owner/architect.

In-wall digital timer, such as Intermatic E101OC, or equal.

Ground Fault Interruptor Circuit

Drawing 16.3
This sample catalogue sheet marks which is the specified luminaire, along with the corresponding symbol.

Used by permission of Lutron

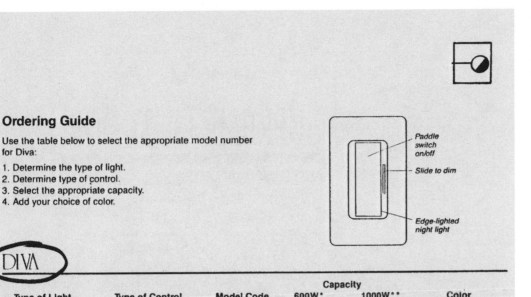

Ordering Guide

Use the table below to select the appropriate model number for Diva:

1. Determine the type of light.
2. Determine type of control.
3. Select the appropriate capacity.
4. Add your choice of color.

Paddle switch on/off

Slide to dim

Edge-lighted night light

DIVA

Type of Light	Type of Control	Model Code	Capacity 600W*	1000W**	Color	
Incandescent	Single-pole	DV-	600P-	10P-	WH	White
	3-way	DV-	603P-	103P-	IV	Ivory
			600VA**	1000VA**	AL	Almond
					GR	Gray
Low-Voltage	Single-pole	DVLV-	600P-	10P-	BR	Brown
	3-way	DVLV-	603P-	103P-	BL	Black

Example: The model number for a low-voltage, 3-way, 600VA Diva dimmer in white is: DVLV-603P-WH.

*Available 4/30/92
**Available 5/29/92

Wallplate Options

Use Diva with Lutron SkyLine™ screwless wallplates or with any standard designer-style wallplate. Both styles are available through your local electrical distributor in colors to match Symphony Series controls. Wallplates are not included with Symphony Series controls.

Worldwide Technical and Sales Assistance

For help with applications, systems layout, or installation, call the toll-free *Lutron Hotline:*
(800) 523-9466 (U.S.A.)
Outside the U.S.A., call (215) 282-3800
FAX: (215) 282-3090

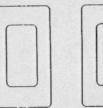

SkyLine screwless wallplate

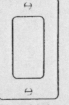

Standard designer-style wallplate

Chart 16.4
This is an additional
sample catalogue page
with the specified luminaire
highlighted and the corresponding symbol added
to the page.

Used by permission of
Justice Design Group

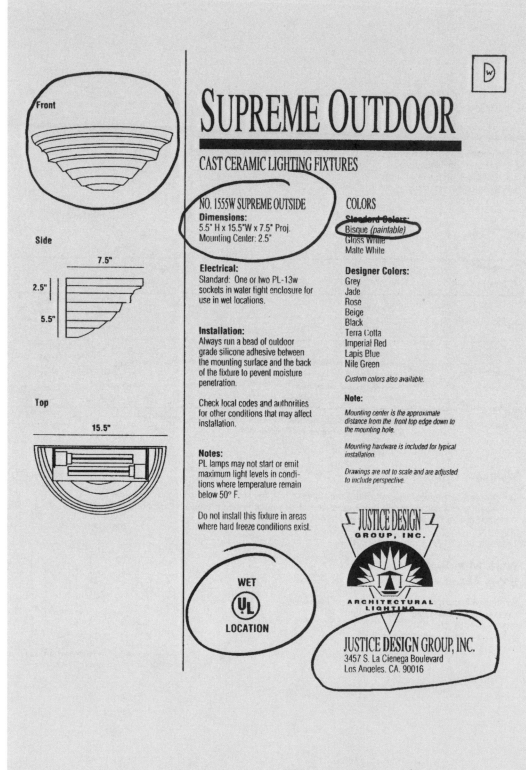

Front

Side

7.5"

2.5"

5.5"

Top

15.5"

SUPREME OUTDOOR

CAST CERAMIC LIGHTING FIXTURES

NO. 1555W SUPREME OUTSIDE

Dimensions:
5.5" H x 15.5"W x 7.5" Proj.
Mounting Center: 2.5"

Electrical:
Standard: One or two PL-13w
sockets in water tight enclosure for
use in wet locations.

Installation:
Always run a bead of outdoor
grade silicone adhesive between
the mounting surface and the back
of the fixture to pevent moisture
penetration.

Check local codes and authorities
for other conditions that may affect
installation.

Notes:
PL lamps may not start or emit
maximum light levels in conditions where temperature remain
below 50° F.

Do not install this fixture in areas
where hard freeze conditions exist.

COLORS

Standard Colors:
Bisque *(paintable)*
Gloss White
Matte White

Designer Colors:
Grey
Jade
Rose
Beige
Black
Terra Cotta
Imperial Red
Lapis Blue
Nile Green

Custom colors also available.

Note:
*Mounting center is the approximate
distance from the front top edge down to
the mounting hole.*

*Mounting hardware is included for typical
installation.*

*Drawings are not to scale and are adjusted
to include perspective.*

WET
(U)L
LOCATION

JUSTICE DESIGN
GROUP, INC.

ARCHITECTURAL
LIGHTING

JUSTICE DESIGN GROUP, INC.
3457 S. La Cienega Boulevard
Los Angeles. CA. 90016

Part C—Lighting Questionaire

This questionaire is a good starting point when meeting with your clients to discuss lighting. Getting these questions answered will help you create a design that suits the project.

To Our Clients—There are many factors that affect the way a room is lighted. By answering the following questions we will be able to give you the lighting design advice that best suits your needs.

1. Room Description: _____

    _____

2. Dimensions: _____

3. New construction or Remodel? (Circle One)
    If it is a remodel project, where are existing lights located? _____

    _____

4. Ceiling height _____
    Sloped or flat (circle one)

5. Is there enough ceiling depth to consider recessed fixtures as an option?
    (Circle one) Yes   No

6. Sky lights (Yes or No) If so, how many and where are they located? _____

    _____

7. If you have sky lights, are they clear, bronze or white? (circle one)

8. Do you have a budget in mind? (circle one)   Yes   No
    If yes, how much? _____

9. What colors are the walls and ceiling painted? _____

    _____

10. Where are the windows located? _____

    _____

11. What do the windows look out onto? _____

    _____

Part D— Condominium Project

You have been asked to design the lighting concepts for a new condo. The clients want you to tell them where to put lights and what type of luminaires you recommend.(See Drawing 16.5)

You are to use the sample list of symbols provided to designate the type of luminaires to be installed. If a symbol is not listed for a luminaire you would like to use, create one and identify it in the blank spaces at the end of the legend.

The reason to start using distinct symbols is to make the lighting plans easy to read. The norm is to use the same symbol to represent almost everything - ceiling-mounted luminaires, wall-mounted luminaires, recessed luminaires, even table lamps. This is confusing to read and can lead to mistakes during the take off for order counts as well as installation.

Background Information

What you know about the clients: your clients are a married couple, 30 years old without children. The walls and ceilings are white and the floors are light oak.

You have already worked out the furniture plan and location of art with these clients. That furniture layout is shown on the plan.

They have enough money from their dual incomes and a home improvement loan to do whatever you recommend. You are not limited by budget.

There will be nine-foot ceilings throughout, except for the living room, which has a pitched ceiling that is 12' high at the apex beam.

There is adequate space available for recessed luminaires.

The condo complex is located in Baton Rouge, Louisiana and their unit is on the first floor.

Throughout the project, make sure you provide adequate lighting for task, ambient, and accent illumination. Don't forget to consider exterior lighting, as well.

Solution—A sample lighting layout can be found in Appendix One.

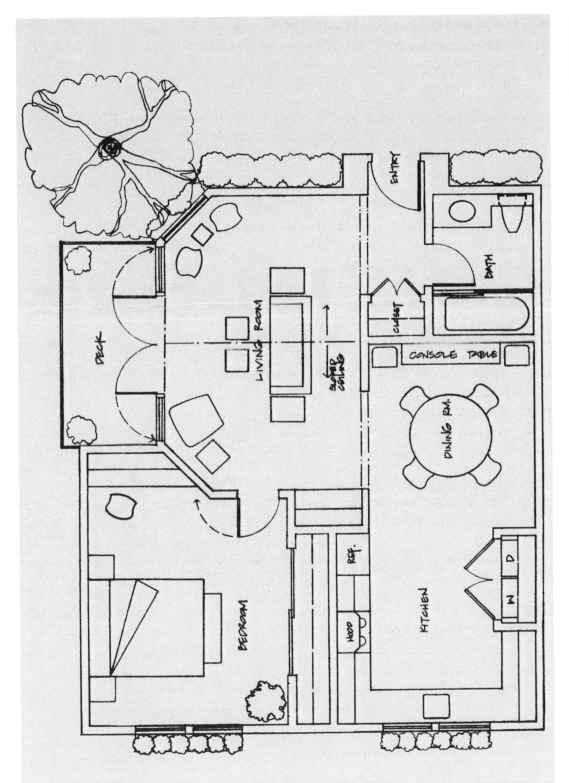

Drawing 16.5
Condo Project
Photocopy this page and try your hand at developing a lighting plan based on the client information given.

Part E: Review Examination - Residential/Landscape Lighting

1. What is meant by layering light? Explain and give examples._____
   _____
   _____

2. What 3 things are lit in a space? Which is the most important? Why?_____
   _____
   _____
   _____
   _____

3. What are color temperature and CRI? How would you use them to choose a lamp?
   _____
   _____
   _____
   _____
   _____
   _____

4. How do finishes and light sources interact in a space? Explain by giving examples?
   _____
   _____
   _____
   _____

5. What are some of the things to consider in specifying a fluorescent luminaire? How do
   they affect your decision? _____
   _____
   _____
   _____
   _____

6. If you were setting the scenes for a 4-scene preset dimmer system, what scenes might
   you use? Why? Describe _____
   _____
   _____
   _____
   _____

7. How does Title 24 affect kitchen and bath lighting in California?_____
   _____
   _____
   _____

8. What are some of the considerations (other than Title 24) used when designing lighting
    for kitchens? Explain. _____
    _____
    _____
    _____
    _____
    _____

9. When lighting landscapes, safety, security and beauty are to be considered. Choose one
    of these and describe some of the techniques used to accomplish this?
    _____
    _____
    _____
    _____

10. Why is it important to know the plant material being used when doing landscape lighting?
    _____
    _____
    _____

11. What are some of the architectural constraints one must deal with in doing lighting
    design? Give an example?_____
    _____
    _____
    _____
    _____

12. Describe the steps you would take in doing a lighting design from start to finish.
    _____
    _____
    _____
    _____
    _____
    _____

13. Color temperature is measured in degrees of _____

14. Do green plants look better under a cooler or a warmer color temperature?_____

15. Name 2 benefits of ambient light?
    1._____
    _____
    _____

2._____
_____
_____

16. Give 2 examples of task lighting in:

An Office: 1._____
             2._____

A Kitchen: 1_____
             2._____

17. How would you improve on the lighting in a dining room that is lighted only by a chan-
    delier?_____
    _____
    _____
    _____
    _____

18. What does the 'MR' stand for in an MR16 lamp? _____
    _____

    (The answer is not "mister")

19. What is the diameter of an MR16 lamp?_____

20. In order to dim a quartz (tungsten halogen) lamp without significantly shortening its
    life, what percentage of time must it be burning at full capacity?_____
    _____

21. Why are clients afraid of fluorescent and what can you do to enlighten them? _____
    _____
    _____
    _____

22. For task lighting at a vanity, where is the best place to locate the luminaires?_____
    _____
    _____
    _____

23. H.I.D. stands for _____

24. Mercury vapor lamps produce what color light?_____
    _____

25. High-pressure sodium lamps produce what color light? _____

26. People tend to look rather unattractive under High Pressure Sodium light. What might look good under this light? _____
_____
_____
_____

27. Is "daylight" a warm color or a cool color temperature?_____

28. What does the 'ER' stand for in a 75ER30 lamp? _____

29. Name two disadvantages of the compact fluorescent.
    1._____
    2._____

30. The three lamp categories are incandescent, fluorescent and H.I.D. Which category does a quartz lamp fit into?_____
_____
_____
_____

31. What are two advantages of fluorescent over incandescent?
    1._____
    _____
    2._____
    _____

32. Describe the "black hole" effect:_____
_____
_____
_____

33. Describe the "moonlighting" effect:_____
_____
_____
_____

34. Why is it necessary to light the clients as well as the art and architecture in a home?
_____
_____
_____
_____

35. If you were glowing from within, what color light would you emit to attract bugs?
_____

36. Give three advantages of low-voltage lighting?
    1._____
    2._____
    3._____

37. Give three disadvantages of low-voltage lighting?
    1._____
    2._____
    3._____

38. True or False? All "pagoda lights" should be sent to Wichita, Kansas to be used as land-ing lights. _____

39. Describe the term "veiling reflection":_____
_____
_____
_____

40. What does not change significantly when you dim fluorescent? _____
_____
_____

41. What is the best way to light a piece of art with a reflective face (such as glass or Plexiglas™)?_____
_____
_____

42. When a client says, "How much light do I need for this room?", list at least three ques-tions you need to ask them about the space:
    1._____
    2._____
    3._____
    4._____
    5._____

43. What is a panic switch and where is it best located? _____
_____

44. What is the most important element to any design project? _____
_____

Glossary

Here is a quick reference guide for terms that are commonly used in lighting design. They have been put in layman's terms so that you can effectively explain them to your clients.

Absorption—Refers to a measure of the amount of light absorbed by an object, instead of being reflected. Dark colored and matte surfaces are least likely to reflect light.

Accent Lighting—Lighting directed at a particular object in order to focus attention upon it.

Ambient Lighting—The soft indirect light that fills the volume of a room with illumination. It softens shadows on people's faces and creates an inviting glow in the room.

Amperage—The amount of electrical current through a conductive source.

Angle of Reflectance—The angle at which a light source hits a specular reflective surface equals the angle at which the resulting glare is reflected back.

ANSI—Five digit numbering system in national use for designing lamp types.

Ballast—Device that transforms electrical energy used by fluorescent, mercury vapor, high and low pressure sodium, or metal halide lamps so the proper amount of power is provided to the lamp.

Beam Spread—The diameter of the pattern of light produced by a lamp or lamp and luminaire together.

Below Grade—Recessed below ground level.

Black Hole—Refers to an opening or window in a room that appears to be empty darkness, especially at night, because there is insufficient illumination at the other side to light up the objects or features framed by the opening.

Bridge System—Two wire, low voltage cable system.

Candle Power—A measure of intensity of light related to lumens.

Cold Cathode—A neon-like electric-discharge light source primarily used for illumination (neon is often used for signage or as an art form). Cold cathode can sometimes be used where fluorescent tubes would be too large or too hard to re-lamp.

Color Rendering Index (CRI)—A scale used to measure how well a lamp illuminates an object's color tones as compared with the color of daylight.

Color Corrected—The addition of phosphors in a lamp to create better CRI

Decorative Luminaire—Luminaire designed to please the eye and provide focal illumination.

Derating—The reduction of the amount of wattage used to prevent overheating. Related to gangin g of dimmers.

Diffusion Filters—Glass lenses used to widen and soften light output.

Dimmer—A control that regulates light levels.

Dimming Ballast—Device used with fluorescent lamps to control the light level. May also apply to H.I.D. sources.

Efficacy—Measurement of the efficiency of a light source.

ETL—An independent testing facility, similer to U.L.

Fade Rate—Rate at which light levels decrease.

Fiberoptics—A illuminating system composed of a lamp source, fiber, and output optics used to remotely light an area or object.

Filters—Glass or metal accessory used to alter beam patterns.

Fish Tape—Mechanical device used to pull wires in tight spaces or conduit.

Fluorescent Lamp—A very energy-efficient type of lamp that produces light through the activation of the phosphor coating on the inside surface of a glass envelope. These lamps come in may shapes, wattages, and colors.

Foot Candle—A measurement of the total light reaching a surface. One lumen falling on one square foot of surface produces the illumination of one foot candle.

Framing Projector—A luminaire that can be adjusted to precisely frame an object with light.

Ganging—Grouping two or more controls in one enclosure.

Glare/Glare Factor—A source of uncomfortably bright light that becomes the focus of attention rather than what it was meant to illuminate.

Halogen—An incandescent lamp containing halogen gas which recycles the tungsten.

Hard Wire—Method of luminaire installation using a junction box.

High Pressure Sodium—An H.I.D. lamp that uses sodium vapor as the light-producing element. It produces a yellow-orange light.

High Intensity Discharge (H.I.D.) Lamp—A category of lamp that emits light through electricity activating pressurized gas in a bulb. Mercury vapor, metal halide, and high pressure sodium lamps are all H.I.D. sources. They are bright and energy-efficient light sources used mainly in exterior environments.

Housing—Enclosure for recessed sockets and trim above the ceiling.

Incandescent Lamp—The traditional type of light bulb that produces light through electricity causing a filament to glow. It is a very inefficient source of illumination.

Junction Box—Enclosure for joining wires behind walls or ceilings.

Kelvin—A measure of color temperature.

Kilowatt—A measurement of electrical usage. A thousand watts equals one kilowatt.

Lamp—What the lighting industry technically calls a light bulb. A glass envelope with gas, coating, or filament that glows when electricity is applied.

Line-Voltage—The 110-120-volt household current, generally standard in North America.

Louver—A metal or plastic accessory used on a luminaire to help prevent glare.

Low-Pressure Sodium—A discharge lamp that uses sodium vapor as the light-producing element. It produces an orange-grey light.

Low-Voltage Lighting—System that uses less than 50-volt current (commonly 12-volt), instead of 110-120-volt, the standard household current. A transformer is used to convert the electrical power to the appropriate voltage.

Lumen—A unit of light power from a light source: the rate at which light falls on one square foot of surface area one foot away from a light source on one candlepower or on candela.

Luminaire—The complete light luminaire with all parts and lamps (bulbs) necessary for positioning and obtaining power supply.

Mercury Lamp—An H.I.D. lamp where the light emission is radiated mainly from mercury. It can be clear, phosphor-coated, or self-ballasted. It produces a bluish light.

Metal Halide Lamp—An H.I.D. lamp where the light comes from radiation from metal halide. It produces the whitest light of the H.I.D. sources.

Mirror Reflector MR16/MR11—Miniature tungsten halogen lamps with a variety of beam spreads and wattages. It is controlled by mirrored facets positioned in the reflector.

Motion Sensor—Control which activates luminaires when movement occurs.

Neon—A glass vacuum tube filled with neon gas and phosphors formed into signs, letters or shapes.

Open Hearth Effect—Lighting that creates the feeling of a glowing fire.

Panic Switch—An on/off switch to activate security lighting, usually located by the bed for emergencies.

PAR Lamps—Lamps (bulbs) with parabolic aluminized reflectors that give exacting beam control. There are a number of beam patterns to choose from, ranging from wide flood to very narrow spot. PAR lamps can be used outdoors due to their thick glass, which holds up in severe weather conditions.

Photo-Pigment Bleaching—Mirror like reflection of the sun on a surface causing glare.

Photosensor—A control device that activates luminaires depending on surrounding light levels.

Planetarium Effect—Too many holes in the ceiling from an over abundance of recessed fixtures.

R Lamp—An incandescent source with a built in reflecting surface.

Reflectance—The ratio of light reflected from a surface

Reflected Ceiling Plan—A lighting plan drawn from the floor looking at the ceiling above.

RLM Reflectors—A luminaire designed to reflect light down and prevent upward light transmission.

Spread Lens—A glass lenses accessory used to diffuse and widen beam patterns.

Stake Lights—Luminaires mounted on a stake to go into the ground or a planter.

Swiss Cheese Effect—Too many holes in the ceiling from an over abundance of recessed fixtures.

Switches—Controls for electrical devices.

Task Lighting—Illumination designed for a work surface so good light, free of shadows and glare, is present.

Timers—Control devices to activate luminaires at set timed intervals.

Transformer—A device which can raise or lower electrical voltage, generally used for low-voltage lights.

Tungsten-Halogen—A tungsten incandescent lamp (bulb) which contains gas and burns hotter and brighter than standard incandescent lamps.

UL—An Independent testing company. Underwriters Laboratory.

Veiling Reflection—A mirror like reflection of a bright source on a shiny surface.

Voltage—A measurement of the pressure of electricity going through a wire.

Voltage Drop—The decrease of light output in fixtures further from the transformer in low voltage lighting systems.

White Light—Usually refers to light with a color temperature between 5000-6250 degrees Kelvin and composed of the whole visible light spectrum. This light allows all colors in the spectrum on an object's surface to be reflected, providing good color-rendering qualities. Daylight is the most commonly referred to source of white light.

Xenon—An inert gas used as a component in certain lamps to produce a cooler color temperature than standard incandescent. It is often used in applications where halogen may normally be specified, because of a longer lamp life.

Index

Appendix One

Resources: How to Use Them

Lighting design is a relatively new and emerging field. A basic lighting course can not begin to cover all aspects of lighting design. It is important that you feel comfortable utilizing all the resources available to you.

Please take advantage of the opportunities to expand your education. There are associations (listed on Page 172) that host classes or seminars around the country on all aspects of lighting design. American Lighting Association (ALA), Illuminative Engineering Society (IES), Designer Lighting Forum and American Society of Interior Designers (ASID) are all open to design professionals.

Take the time to visit your local lighting showroom; investigate and learn how to use their lighting labs. Most showrooms now have recessed, track and outdoor labs of some sort, along with full displays of dimming options.

Consult with and try to work with a particular individual you feel is knowledgeable and reliable. Establish a raport so you feel comfortable asking questions and know that the answers will be correct. Ask for their credentials.

American Lighting Association has a Certified Lighting Consultant program and IES has a program in place for Certificates of Technical Knowledge.

Please check with your local building departments to verify local and state code requirements. Most have handbooks available for your library. It is far better to ask questions than to have to replace a ceiling because you specified the wrong placement of recessed fixtures.

In many cases, it might be better to seek the advice of a professional lighing consultant. While scarce in mid-America, the east and west coasts as well as Chicago and Dallas, have a number of highly qualified experts.

Interior Design Market Centers put on seminars in conjunction with their annual, or twice yearly, market events.

You can never know too much, so take advantage of this emerging field. New and exciting lamps are rapidly being developed and keeping up is as important as new fashion styles or colors.

Illuminating Engineering Society of North America (IES)
345 E. 47th Street
New York, NY 10017

Designers Lighting Forum (DLF)
Contact your local IES Chapter

American Lighting Association (ALA)
World Trade Center, SUite 10046
P.O. Box 580168
2050 Stemmons Freeway
Dallas, TX 75258-0168

American Society of Interior Designers (ASID)
608 Massachusetts Ave, N.E.
Washington, D.C. 20002-6006

Appendix Two

SAMPLE ANSWERS AND LAYOUTS

Sample Answers to Review Examination

1. What is meant by layering light? Explain and give examples._____
   The use of a combination of luminaires to create a cohesive overall design. A blend of task, ambient, accent, and decorative sources.

2. What 3 things are lit in a space? Which is the most important? Why?_____
   People, art, architecture. People are the most important. They need to feel comfortable, look their best, and have access to controllable, functional illumination.

3. What are color temperature and CRI? How would you use them to choose a lamp?
   Color temperature measures in degrees of Kelvin the color quality of a lamp. The CRI (color rendering index) compares the lamp source to daylight and its ability to mimic daylight.

4. How do finishes and light sources interact in a space? Explain by giving examples?
   Lamps and luminaires must be selected based on the finishes in a particular space. Dark surfaces absorb light, while light colored surfaces reflect light, so deeply hued rooms need more light.

5. What are some of the things to consider in specifying a fluorescent luminaire? How do they affect your decision? _____
   Consider using an electronic ballast, so that the luminaire is as quiet as possible. A dimming ballast would allow for varying light levels.

6. If you were setting the scenes for a 4-scene preset dimmer system, what scenes might you use? Why? Describe _____
   Scene One - normal - everyday setting
   Scene Two - pass through (a low light level - enough to see)
   Scene three - party - a little more accent and less ambient
   Scene four - clean-up - all lights up to full

7. How does Title 24 affect kitchen and bath lighting in California?_____
   The general illumination in new construction or remodel that exceeds
   50% must use fluorescent as the general illumination and be on the
   first switch as you enter the room.
   _____

8. What are some of the considerations (other than Title 24) used when designing lighting for kitchens? Explain. _____
   Provide good general illumination to help reduce shadowing.
   Locate task lighting between your head and your work surface, such
   as under-cabinet lighting.
   If entertaining includes the kitchen area, make the lighting as
   comfortable and controllable as the rest of the public spaces.

9. When lighting landscapes, safety, security and beauty are to be considered. Choose one of these and describe some of the techniques used to accomplish this?
   Where beauty is concerned, work on lighting the plantings, without
   drawing attention to the luminaires themselves. Let the decorative
   lanterns just be a glow and give the illusion that they are providing
   the light.

10. Why is it important to know the plant material being used when doing landscape lighting?
    Trees that lose their leaves may be lighted differently than ever-
    greens. Fast-growing plants may overtake a luminaire; extra lengths
    of cable would allow the luminaire to be moved as the plants mature.

11. What are some of the architectural constraints one must deal with in doing lighting design? Give an example?_____
    Existing structures can take advantage of luminaires made especially
    for remodel, but make sure that there is enough ceiling depth to
    accommodate a recessed can.
    _____

12. Describe the steps you would take in doing a lighting design from start to finish.
    1. Sit with the client to talk about their needs and their budget.
    2. Draw up a furniture plan.

3. *Draw up a preliminary lighting plan and specifications to review with the clients.*
4. *Draw up the finished plan and specifications.*
5. *Review with the installing contractor.*

13. Color temperature is measured in degrees of __*Kelvin*__

14. Do green plants look better under a cooler or a warmer color temperature? *cooler*

15. Name 2 benefits of ambient light?
    1. *Softens shadows on people's faces so that they look their best.*

    2. *Fills the volume of a space to make it seem larger and more inviting.*

16. Give 2 examples of task lighting in:

    An Office: 1. *A desk lamp directed towards the keyboard of a computer*
                 2. *A ceiling-mounted luminaire in the supply room.*

    A Kitchen: 1. *Lighting under the overhead cabinets*
                 2. *A ceiling-mounted luminaire in the pantry*

17. How would you improve on the lighting in a dining room that is lighted only by a chandelier? _____
    Add a recessed adjustable fixture on either side of the chandelier to cross-illuminate the table. Install wall sconces or torchieres to provide the much needed ambient illumination.

18. What does the 'MR' stand for in an MR16 lamp? _____
    Mirror Reflector
    (The answer is not "mister")

19. What is the diameter of an MR16 lamp? *2" (16 eighths of an inch)*

20. In order to dim a quartz (tungsten halogen) lamp without significantly shortening its life, what percentage of time must it be burning at full capacity? *20%*

21. Why are clients afraid of fluorescent and what can you do to enlighten them? _____
    Fluorescents come in many wonderful colors.
    Fluorescents give 3 to 5 times more light than comparable wattage incandescents.
    Fluorescents last 10 to 30 times longer than standard incandescents.

22. For task lighting at a vanity, where is the best place to locate the luminaires? _____
    Flanking the mirror, mounted at eye level.

23. H.I.D. stands for *High intensity discharge*

24. Mercury vapor lamps produce what color light? *bluish-green*

25. High-pressure sodium lamps produce what color light? *yellow-orange*

26. People tend to look rather unattractive under High Pressure Sodium light. What might look good under this light? _____
    Brick, sandstone, the Golden Gate Bridge

27. Is "daylight" a warm color or a cool color temperature? *Cool*

28. What does the 'ER' stand for in a 75ER30 lamp? *Ellipsoidal reflector*

29. Name two disadvantages of the compact fluorescent.
    1. *Some hum.*
    2. *Some don't have rapid-start ballasts.*

30. The three lamp categories are incandescent, fluorescent and H.I.D. Which category does a quartz lamp fit into? _____
    Incandescent

31. What are two advantages of fluorescent over incandescent?
    1. *Longer life*

    2. *Greater variety of colors*

32. Describe the "black hole" effect:_____
When a lighting design has neglected to light the exterior spaces, windows can become "black holes" at night. Exterior lighting helps eliminate this problem.

33. Describe the "moonlighting" effect:_____
Positioning exterior fixtures so that they filter through the branches of the trees creates an effect similar to light cast by the moon.

34. Why is it necessary to light the clients as well as the art and architecture in a home?
The space needs to be humanized, so that the homeowners can feel comfortable in the space. Lighting the people in a space helps them look and feel their best.

35. If you were glowing from within, what color light would you emit to attract bugs?
Blue-white

36. Give three advantages of low-voltage lighting?
1. *Compact lamps*
2. *Tight beam spreads*
3. *Energy-efficient*

37. Give three disadvantages of low-voltage lighting?
1. *Hum*
2. *Voltage drop*
3. *Limited wattage*

38. True or False? All "pagoda lights" should be sent to Wichita, Kansas to be used as landing lights. *True*

39. Describe the term "veiling reflection": *The mirror-like reflection of a light source on a shiny surface when light comes from the ceiling directly in front of you, hitting the paper at such an angle that the glare is reflected directly into your eyes, as if trying to read through a veil.*

40. What does not change significantly when you dim fluorescent? _____
The color temperature

41. What is the best way to light a piece of art with a reflective face (such as glass or Plexiglas™)? *Cross-illumination is best, using two luminaires. The one on the right is directed towards the left side of the art and vice-versa.*

42. When a client says, "How much light do I need for this room?", list at least three questions you need to ask them about the space:
    1. _What color will the walls and ceilings be painted?_
    2. _Will they be doing tasks, such as reading, in the space?_
    3. _Will they be entertaining in the space?_
    4. _Are there skylights?_
    5. _What is the finish or trim?_

43. What is a panic switch and where is it best located? _____
    A switch located next to the bed to turn on exterior lights.

44. What is the most important element to any design project? _Lighting is great, but money ends up being the true most important element._

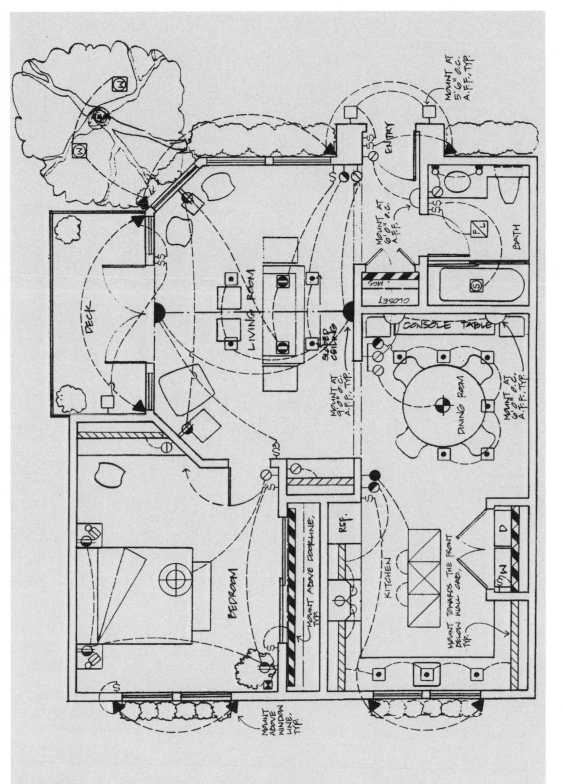

Sample lighting layout for the condo project in Chapter 16. Check these symbols against the ones listed in Chart 16.2 to get a better understanding of what has been designed.

Appendix Three

A SAMPLING OF LUMINAIRE MANUFACTURERS

AAMSCO Manufacturing Inc.
  (vertical vanity lights)
  15-17 Brook St.
  P.O. Box 15119
  Jersey City, NJ 07305
  201/434-0722
  FAX: 201/434-8535

Arroyo Craftsman Lighting, Inc.
  (Period exterior luminaires)
  44509 Little John Street
  Baldwin Park, CA 91706
  818/960-9411
  FAX: 818/960-9521

ALKCO
  (Under-cabinet task lights)
  11500 West Melrose Street
  Franklin Park, IL 60131
  312/451-0700

Artimede Inc.
  (Italian pendants and wall sconces)
  1980 New Highway
  Farmingdale, NY 11735
  516/694-9292
  FAX: 516/694-9275

Banci
  (Traditional decorative luminaires with
  hidden indirect halogen lamps)
  Rep: Casella Lighting
  111 Rhode Island St.
  San Francisco, CA 94103
  415/626-9600
  FAX: 415/626-4539

Bega/US
  (Exterior step lights and fixtures)
  1005 Mark Avenue
  Carpinteria, CA 93013
  805/684-0533
  FAX: 805/684-6682

Norbert Belfer Lighting Mfg. Co.
  (Compact fluorescent strip lights,
  halogen strip lights)
  1703 Valley Road
  Ocean, NJ 07712
  908/493-2666
  FAX: 908/493-2941

B-K Lighting, Inc.
  (Recessed step lights)
  2720 N. Grove Industrial Drive
  Fresno, CA 93727
  209/255-5300
  FAX: 209/255-2053

Boyd Lighting Company
(Decorative wall sconces, torchieres, pendant luminaires)
56 12th Street
San Francisco, CA 94103
415/431-4300
FAX:415/431-8603

Capri
(Recessed light luminaires)
6430 E. Slauson Ave.
Los Angeles, CA 90040
213/726-1800
FAX: 800-444-1886

Casella
(Mini/torchieres)
111 Rhode Island St.
San Francisco, CA 94103
415/626-9600
FAX: 415/626-4539

CSL Lighting
("Invizilite", miniature recessed wall washers)
27615 Avenue Hopkins
Valencia, CA 91355
805/257-4155
FAX: 805/257-1554

Exciting Lighting
(Light sculpture)
14 E. Sir Francis Drake Blvd.
Larkspur, CA 94939
415/925-0840
FAX: 415/925-1305

Fiberstar
(Fiber optics)
47456 Fremont Blvd.
Fremont, CA 94538
800/FBR-STRS or 415/490-0719
FAX: 415/490-3247

Georgian Art
(Decorative exterior lanterns)
P.O. Box 325
Lawrenceville, GA 30246
404/963-6221
404/963-6225

Hadco
(Exterior luminaires)
P.O. Box 128
Littlestown, PA 17340
717/359-7131
FAX: 717/359-9289

Halo
(Recessed light luminaires, daylight blue filters)
400 Busse Road
Elk Grove Village, IL 60007
708/956-8400
708/956-1537

Hubbell
(120-volt L-andscape light luminaires)
2000 Electric Way
Christiansburg, VA 24073
703/382-6111
FAX: 703/382-1526

Imperial Bronzelite
P.O. Box 606
San Marcos, TX 78667-0606
512-392-8957
FAX: 512-353-5822

Juno
(Track and recessed luminaires)
2001 S. Mt. Prospect Road
P.O. Box 5065
Des Plaines, IL 60017-5065
708/827-9880
FAX: 708-827-2925

Justice Design Group
  (Bisque wall sconces)
  3457 La Cienega Blvd.
  Los Angeles, CA 90016
  310/836-9575
  FAX: 310/836-0204

Kim Lighting
  (Landscape light luminaires)
  16555 East Gale Avenue
  Industry, CA 91749
  818/968-5666
  FAX: 818/369-2695

Koch and Lowy
  (Vodka pendant luminaires,
  wall sconces, torchieres)
  21-24 39th Ave.
  Long Island City, NY 11101
  718/786-3520
  FAX: 718/937-7968

Lightolier
  (Recessed low-voltage adjustable
  luminaires, decorative luminaires)
  100 Lighting Way
  Secaucus, NJ 07096
  201/864-3000
  FAX: 201/864-9478

Loran/Nightscaping
  (12-volt Landscape lighting luminaires)
  1705 East Colton Avenue
  Redlands, CA 92373
  909/794-2121
  FAX: 909/794-2121

Lucifer
  (Miniature low-voltage cove light
  luminaires)
  414 Live Oak Street
  San Antonio, TX 78202
  210/227-7329
  FAX: 210/227-4967

Lutron
  (Low-voltage; line-voltage; and solid-
  state fluorescent dimming systems)
  7200 Suter Road
  Coopersburg, PA 18036
  800/523-9466
  FAX: 215/282-3090

Nova
  (Opaque pendant luminaires,
  wall sconces, honeycomb louvers)
  999 Montague Avenue
  San Leandro, CA 94577
  510/357-0171
  FAX: 510/357-3832

Phoenix Day
  (Decorative plaster wall sconces,
  torchieres, vanity luminaires, opaque
  pendant luminaires, etc.)
  1355 Donner Avenue
  San Francisco, CA 94124
  415/822-4414
  415/822-3987

Poulsen
  (Modern exterior luminaires)
  5407 NW 163rd Street
  Miami, FL 33014-1009
  305/625-1009
  FAX: 305/625-1213

Prescolite
  (low-voltage recessed adjustable
  luminaires)
  1251 Doolittle Dr.
  San Leandro, CA 94577
  510/562-3500

Shaper Lighting
(custom wall luminaires and pendants,
low-voltage exterior lighting)
1141 Marina Way South
Richmond, CA 94804-3742
510/234-2370
FAX: 510/234-2371

Starfire
("Xenflex" - Xenon strip lighting)
317 St. Pauls Ave.
Jersey City, NJ 07306
800-443-8823 FAX: 201-656-0666

Sylvan Designs, Inc.
(Light buds: low voltage miniature
halogen linear lighting)
8921 Quartz Avenue
Northridge, CA 91324
818-998-6868 FAX: 818-998-7241

Osram Sylvania
("Designer 16" - 12-volt MR16)
GTEJ Products Corp.
Sylvania Lighting Center
100 Endicott Street
Danvers, MA 01923
508-777-1900

Visa Lighting
(Pendants, wall sconces,
exterior lighting)
8600 West Bradley Road
Milwaukee, WI 53224
414-354-6600
Fax: 4414-354-7436
Specifier hot line: 1-800-788-VISA

Wellmade
(Fluorescent luminaires)
860 81st Avenue
Oakland, CA 94621
510-562-1878
FAX: 510-632-7865

Wendelighting
(framing projectors)
2445 North Naomi Street
Burbank, CA 91504
818/955-8066
800/528-0101
FAX: 818/848-0674

Zelco
(European-style decorative luminaires,
recessed wall sconces)
630 S. Columbus Ave. C.S. #4445
Mt. Vernon, NY 10550
800-431-2486
914/699-6230
FAX: 914/699-7082